Chapter 1: Introduction to Endorphins

What are Endorphins?

Endorphins are a group of naturally occurring peptides in the brain and nervous system that act as neurotransmitters, meaning they transmit signals between nerve cells. These powerful chemicals are often referred to as the body's "natural painkillers" because of their role in alleviating pain and promoting a sense of well-being. The word "endorphin" is derived from two parts: "endo," meaning internal, and "orphin," which is short for morphine, reflecting their morphine-like effects on the body.

Despite being most famous for their association with pain relief, endorphins also play crucial roles in mood regulation, stress management, and enhancing overall mental and physical well-being. Their primary function is to create a positive feeling in the body, often linked with activities that promote physical exertion, pleasure, and social bonding.

Their Role in the Human Body

Endorphins are produced by the central nervous system (CNS), particularly in the brain and spinal cord. Their functions extend beyond just mood elevation and pain relief. These molecules are involved in regulating a wide variety of bodily processes, including:

- **Pain management:** Endorphins help to reduce pain perception by binding to opioid receptors in the brain, blocking pain signals from reaching the conscious mind.

- **Stress response:** When under stress, the body releases endorphins to help manage the physiological impact of stress and maintain homeostasis.

- **Mood enhancement:** By binding to specific receptors in the brain, endorphins trigger positive feelings and emotional well-being.

- **Exercise response:** Physical activity stimulates the production of endorphins, contributing to the well-known "runner's high" and promoting recovery and endurance.

Through these mechanisms, endorphins are essential in maintaining both physical health and emotional balance.

Types of Endorphins and Their Functions

There are several types of endorphins in the body, but the most well-known and studied are:

- **Beta-endorphins:** The most potent and widely distributed endorphins in the body, beta-endorphins are involved in pain relief and feelings of euphoria. They are predominantly produced in the pituitary gland and play a significant role in the body's response to exercise, stress, and injury.

- **Alpha-endorphins:** While not as potent as beta-endorphins, alpha-endorphins still play a role in modulating pain and mood. These are produced in smaller amounts and can be more localized in certain regions of the brain.

- **Enkephalins and Dynorphins:** These are smaller peptides that function similarly to endorphins, acting on opioid receptors to modulate pain, mood, and stress responses. They are often involved in short-term stress responses and can have varying effects based on receptor activation.

Each of these endorphins works in different ways to provide pain relief, induce pleasure, and regulate various bodily systems.

Historical Background and Scientific Discovery

The discovery of endorphins dates back to the 1970s, when researchers first identified the existence of naturally occurring substances in the brain that could bind to opioid receptors. This discovery was revolutionary, as it helped explain why the body experienced pain relief in response to exercise, stress, or injury without the need for external painkillers.

The term "endorphin" was coined by scientists from the University of California, Los Angeles (UCLA), specifically Dr. Solomon Snyder and Dr. Candace Pert. Their groundbreaking work demonstrated how these endogenous opioid peptides mimicked the effects of morphine and other opioid drugs, but without the negative side effects, such as addiction or overdose risk.

The 1970s research on endorphins opened the door to further studies on their involvement in mood regulation, stress management, and overall health. Since then, scientists have continued to explore how endorphins interact with other hormones and neurotransmitters to regulate mood, behavior, and cognitive function. Endorphin research has provided valuable insight into areas such as pain management, mental health, athletic performance, and even the aging process.

Today, the understanding of endorphins has expanded beyond the lab, with practical applications in both medical and wellness contexts. Strategies for optimizing endorphin production—such as exercise, nutrition, mindfulness, and stress reduction—are increasingly recognized as essential tools for enhancing both physical and mental health.

Why Master Endorphin Synthesis, Production, and Availability?

While the body naturally produces endorphins in response to certain stimuli, many people may not be harnessing their full potential. By understanding how endorphins work and how to optimize their production, individuals can significantly improve their quality of life. Proper endorphin regulation can provide benefits such as:

- **Enhanced mood and emotional stability**
- **Improved stress resilience and coping mechanisms**
- **Effective pain management without reliance on medication**
- **Better physical performance, recovery, and endurance**
- **Greater mental clarity, focus, and motivation**

This book will explore how to master the processes of endorphin synthesis, production, and availability through natural and practical means. Whether through exercise, diet, sleep, or mindfulness practices, the aim is to empower readers to boost their endorphin levels, leading to a more fulfilled and balanced life. Understanding and optimizing endorphin function is an investment in both mental and physical well-being that can have lifelong benefits.

In the following chapters, we will dive deeper into the science of endorphin synthesis, explore practical strategies for boosting production, and address common challenges and misconceptions. By the end of this book, you will be equipped with the knowledge to harness the power of endorphins, enhancing your overall well-being and achieving a state of optimal health and happiness.

Chapter 2: The Biochemistry of Endorphin Synthesis

Understanding how endorphins are synthesized, produced, and made available in the body is key to mastering their benefits. In this chapter, we will explore the biochemical processes that give rise to these potent molecules, the pathways through which they are synthesized, and how their availability can be regulated.

Endorphins: Peptide Hormones and Neurotransmitters

Endorphins are classified as **peptide hormones** and **neurotransmitters**. Peptides are chains of amino acids that are joined by peptide bonds. When released, endorphins can bind to opioid receptors in the brain, spinal cord, and peripheral tissues, where they exert their effects on mood, pain, and various other physiological functions.

As neurotransmitters, endorphins transmit chemical signals between nerve cells, playing a crucial role in brain communication. Their ability to reduce pain perception, regulate stress, and enhance pleasure makes them vital for emotional and physical health.

The synthesis of endorphins, like other peptide hormones, begins with a precursor protein. This precursor is then broken down into smaller peptides, including endorphins, that can interact with various receptors in the body.

Biosynthesis Pathways of Endorphins

Endorphin synthesis primarily occurs in the **pituitary gland**, which is located at the base of the brain. The process begins with the production of **pro-opiomelanocortin (POMC)**, a precursor protein. POMC is produced in response to certain stimuli, such as stress, physical activity, or pain.

Once POMC is synthesized, it undergoes enzymatic cleavage, resulting in the formation of several biologically active peptides, including **beta-endorphins, alpha-endorphins,** and **enkephalins**. These peptides have distinct functions, but they all share the ability to bind to the opioid receptors and exert their effects on the body.

- **Beta-endorphins**: The most studied and potent of the endorphins, these are primarily involved in pain relief and the feeling of euphoria. They are released in response to physical stressors like exercise and injury.

- **Enkephalins**: These smaller peptides are more localized and involved in short-term pain relief and mood regulation. They play a critical role in reducing acute pain and stress.

- **Alpha-endorphins**: While less potent than beta-endorphins, alpha-endorphins are still important in regulating mood and stress responses.

These peptide hormones can travel through the bloodstream to various organs and tissues, exerting their influence through specific opioid receptors in different parts of the body.

The Role of the Pituitary Gland and Hypothalamus

The **pituitary gland** is often referred to as the "master gland" because it controls the release of several important hormones that regulate growth, metabolism, and stress responses. Endorphins are synthesized and released from the pituitary gland in response to various signals from the hypothalamus.

The **hypothalamus** plays a central role in regulating body temperature, hunger, and emotional responses. When the body is under physical or emotional stress, the hypothalamus triggers the release of POMC from the pituitary gland, which then gets converted into endorphins. This chain of events helps the body cope with stressors, including physical injury, emotional strain, and exercise.

In response to the stimuli, the hypothalamus activates the **hypothalamic-pituitary-adrenal (HPA) axis**, a complex network of glands and hormones that regulate the body's reaction to stress. The activation of this axis not only promotes the release of endorphins but also stimulates the production of other stress-related hormones, such as cortisol, which plays a role in managing the body's response to physical and psychological stress.

Understanding Opioid Receptors and Their Mechanisms

Endorphins exert their effects by binding to **opioid receptors** in the brain and body. These receptors are part of a complex system that includes **mu (μ)**, **delta (δ)**, and **kappa (κ)** receptors, each of which responds to different endogenous peptides and exogenous substances, such as drugs like morphine and heroin.

- **Mu receptors** are the most prominent in the brain and are primarily involved in pain relief, euphoria, and other rewarding effects. Beta-endorphins predominantly interact with these receptors.

- **Delta receptors** play a role in modulating mood and have been linked to the emotional effects of endorphins.

- **Kappa receptors** are involved in pain perception and have a more complex role in regulating stress responses.

When endorphins bind to these receptors, they trigger a cascade of biochemical events inside the cell that can lead to the inhibition of pain signals, changes in emotional state, or increased feelings of well-being. The activation of these receptors can also produce a feeling of euphoria, which is why activities such as exercise, laughter, and certain social interactions that stimulate endorphin production are associated with an improved mood.

Understanding the mechanisms of opioid receptors is vital for unlocking the therapeutic potential of endorphins. Researchers are constantly exploring how these receptors interact with endorphins, both naturally occurring and synthetic, to develop more effective treatments for pain, mood disorders, and stress-related conditions.

The Interplay Between Endorphins and Other Neurotransmitters

Endorphins do not function in isolation; they interact with a variety of other neurotransmitters to produce a balanced response in the body. One of the most significant interactions occurs between endorphins and **dopamine**, a neurotransmitter associated with reward and motivation.

When endorphins activate opioid receptors, they can also indirectly boost dopamine production, which enhances feelings of pleasure and reward. This is why activities like exercise or eating enjoyable foods can result in both the release of endorphins and the subsequent dopamine surge, leading to a reinforced feeling of happiness or satisfaction.

Similarly, endorphins interact with **serotonin**, a neurotransmitter involved in regulating mood, anxiety, and happiness. The release of endorphins can modulate serotonin levels, leading to improved mood and a sense of relaxation. This interplay between endorphins, dopamine, and serotonin is a key part of the body's reward system, contributing to overall emotional balance and well-being.

Regulating Endorphin Availability

While endorphins are produced naturally by the body, their availability can be influenced by various factors, such as genetics, lifestyle choices, and environmental conditions. For example, regular physical activity, a healthy diet, and stress management techniques can stimulate endorphin production and ensure that these beneficial chemicals are readily available when needed.

The availability of endorphins can also be influenced by **endocrine signals**, such as those released by the adrenal glands in response to stress. Chronic stress or prolonged periods of discomfort can alter the balance of endorphins and other neurotransmitters, leading to a decrease in endorphin production and, in some cases, a phenomenon known as **endorphin resistance**, where the body's opioid receptors become less sensitive to endorphins.

By understanding the biochemistry of endorphins and the factors that regulate their synthesis and availability, individuals can take practical steps to optimize their endorphin levels. Whether through exercise, nutrition, sleep, or stress management, the goal is to maintain an environment in the body where endorphins are produced efficiently and are available to promote well-being.

Chapter 3: Endorphins and Pain Management

One of the most profound roles that endorphins play in the human body is their involvement in **pain management**. As natural analgesics, endorphins help to mitigate the sensation of pain and promote a sense of well-being. This chapter explores how endorphins function as painkillers, their relationship with pain receptors, and the clinical applications of endorphins in managing pain. Understanding how to enhance endorphin production can be a powerful tool in pain management and recovery.

Endorphins as Natural Painkillers

Endorphins are often referred to as the body's **natural painkillers**. This is because, like morphine and other opioid-based medications, they bind to opioid receptors in the brain and spinal cord to produce pain-relieving effects. However, unlike synthetic opioids, endorphins are not associated with the harmful side effects of addiction or overdose, making them a safer, naturally occurring alternative to pharmacological pain relief.

When the body experiences pain, whether from injury, inflammation, or other causes, it signals the nervous system to produce endorphins. These endorphins then travel through the bloodstream, reaching various parts of the brain and spinal cord where they interact with opioid receptors. Upon binding to these receptors, endorphins trigger a cascade of biochemical signals that reduce the sensation of pain. This mechanism is one of the primary ways the body helps to manage acute pain and maintain homeostasis.

In addition to providing pain relief, endorphins can also induce a **sense of euphoria**, often referred to as a "high." This feeling is part of the body's reward system, which is designed to encourage behaviors that promote survival and well-being, such as exercise, social bonding, and engaging in pleasurable activities.

The Relationship Between Endorphins and Pain Receptors

The pain-relieving effects of endorphins are largely mediated through their interaction with **opioid receptors** in the brain and spinal cord. These receptors are part of the body's endogenous pain control system, often referred to as the **opioid system**.

- **Mu (μ) receptors**: These are the primary receptors involved in the analgesic effects of endorphins. When activated, they reduce pain signals and promote feelings of pleasure or euphoria.
- **Delta (δ) receptors**: These receptors contribute to mood regulation and pain modulation but are less potent than mu receptors in the pain relief process.
- **Kappa (κ) receptors**: Kappa receptors have a more complex role in pain management, particularly in regulating stress and emotional responses to pain.

When endorphins bind to these receptors, they inhibit the transmission of pain signals along the spinal cord and to the brain. This can result in reduced perception of pain, making endorphins highly effective in both acute and chronic pain situations.

Endorphins do not simply block pain in a direct way; they can also enhance the body's ability to tolerate pain and modify the emotional response to it. For instance, someone who experiences pain during a workout may not feel the intensity of the discomfort as acutely due to the release of endorphins. Similarly, those who undergo emotional pain—such as stress, anxiety, or grief—may find that endorphins help them cope by reducing the emotional weight of the experience.

How Endorphins Mitigate Physical Pain

The mechanism by which endorphins mitigate physical pain involves several biological processes:

1. **Pain Signal Suppression**: When the body experiences injury or inflammation, pain signals are sent to the brain via sensory neurons. Endorphins bind to opioid receptors located on these pain-transmitting neurons, reducing their ability to send pain signals to the brain.

2. **Descending Pain Modulation**: In addition to blocking pain signals, endorphins activate descending pathways from the brain that can dampen pain signals coming from the spinal cord. This means that endorphins not only inhibit pain at the site of injury but also enhance the brain's ability to ignore or downplay pain signals.

3. **Reduction of Inflammatory Response**: Endorphins have been shown to modulate inflammation, which is often a major source of pain. By reducing the production of pro-inflammatory cytokines, endorphins can lessen the pain associated with conditions like arthritis, fibromyalgia, and other inflammatory disorders.

4. **Psychological Pain Relief**: Beyond physical pain, endorphins play a role in managing **psychological pain**. Mental and emotional stress often exacerbate physical pain, but the release of endorphins helps to reduce feelings of anxiety and distress, which in turn lessens the perceived intensity of pain.

The powerful combination of these effects—physical pain relief, psychological stress reduction, and enhanced mood—makes endorphins an essential player in the body's pain management toolkit.

Clinical Applications in Pain Management

Understanding the role of endorphins in pain relief has significant clinical implications. Several pain management strategies leverage the body's ability to produce endorphins, either by stimulating their release or by mimicking their effects.

Exercise as Pain Relief

Exercise and Chronic Pain

fibromyalgia

2. **Acupuncture and Endorphin Release**: Acupuncture, an ancient Chinese practice that involves inserting thin needles into specific points on the body, has been shown to stimulate endorphin production. It is often used as a complementary therapy for pain management, particularly in conditions like **back pain**, **headaches**, and **joint pain**.

3. **Massage Therapy**: Another therapy that boosts endorphin levels is **massage**. By stimulating the body's sensory nerves and improving circulation, massage helps to release endorphins, reduce muscle tension, and alleviate pain.

4. **Pharmacological Alternatives**: In some cases, medications that mimic or enhance the effects of endorphins, such as **opioid drugs**, are used in the management of severe pain. However, because of the risks of dependency and other side effects, non-pharmacological methods like exercise, acupuncture, and mindfulness are becoming more widely accepted as safer alternatives.

5. **Mind-Body Techniques**: Practices like **meditation**, **mindfulness**, and **deep breathing** can also trigger the release of endorphins. These techniques are often used to manage pain associated with chronic conditions like **migraines, stress-induced pain**, and **lower back pain**.

6. **Cognitive Behavioral Therapy (CBT)**: For chronic pain sufferers, therapy that involves learning how to modify pain perceptions and emotional responses can enhance endorphin release. CBT can help patients reframe their pain experience and reduce the emotional weight of pain, which enhances endorphin activity and improves pain tolerance.

Conclusion: Harnessing the Power of Endorphins for Pain Management

Endorphins are powerful, natural painkillers that can be leveraged for both acute and chronic pain relief. Their ability to block pain signals, reduce inflammation, and improve emotional resilience makes them a cornerstone of the body's pain management system. By understanding how to stimulate endorphin production through exercise, mindfulness, and other techniques, individuals can effectively manage pain without relying solely on pharmaceutical interventions. Mastering endorphins is not just about pain relief—it is about optimizing the body's natural processes to enhance well-being, resilience, and quality of life.

In the next chapters, we will delve deeper into how lifestyle factors, such as physical activity and nutrition, can further promote endorphin production and availability, enabling us to manage pain, enhance mood, and support long-term health and vitality.

Chapter 4: The Role of Exercise in Endorphin Production

Physical activity is one of the most powerful and natural ways to stimulate the production of endorphins, the body's internal mood boosters. Whether through aerobic exercise, strength training, or even yoga, regular physical activity has profound effects on mental and physical health. In this chapter, we will explore how exercise stimulates endorphin synthesis, the physiological mechanisms behind the "runner's high," and how different types of exercise can be optimized to boost endorphin production. We will also discuss the role of exercise in improving overall well-being and offer practical tips for incorporating exercise into your life to maximize endorphin production.

How Physical Activity Stimulates Endorphin Synthesis

Exercise serves as a powerful trigger for the release of endorphins. When we engage in physical activity, the body perceives stress—whether from exertion, cardiovascular strain, or muscle fatigue. In response, the body produces and releases endorphins to help alleviate the discomfort associated with physical stress. This natural biochemical response not only aids in pain relief but also produces feelings of well-being and euphoria.

Endorphin release during exercise is influenced by several factors:

- **Exercise Intensity**: Higher-intensity exercise, particularly activities that push the body beyond its typical comfort zone, tends to trigger greater endorphin production. Activities like running, cycling, swimming, and weightlifting stimulate the production of endorphins in response to the physical strain they place on the body.

- **Duration**: The longer the duration of physical exertion, the more endorphins the body typically produces. This is why endurance athletes—such as marathon runners or long-distance cyclists—often experience prolonged feelings of euphoria after their workouts.

- **Type of Exercise**: Both aerobic and anaerobic exercises can stimulate endorphin release. While aerobic exercises such as running, rowing, and cycling are known for their strong endorphin-boosting effects, anaerobic exercises like strength training and high-intensity interval training (HIIT) can also trigger significant endorphin release.

The "Runner's High" Phenomenon

One of the most widely recognized experiences associated with endorphin release during exercise is the phenomenon known as the **"runner's high."** This is the intense feeling of euphoria, increased happiness, and mental clarity that some people experience after engaging in sustained, moderate-to-high-intensity aerobic exercise.

The runner's high is believed to occur when endorphins bind to the **mu-opioid receptors** in the brain, producing feelings of happiness and well-being. However, the phenomenon is not limited to running. Cyclists, swimmers, and those who engage in any form of prolonged, moderate-to-intense exercise can experience a similar effect.

Interestingly, the runner's high is not just limited to exercise's physical effects. Researchers suggest that **psychological factors**, such as the sense of accomplishment or overcoming mental barriers, also contribute to the feelings of euphoria. This combination of physiological and psychological responses enhances both mental clarity and emotional resilience, which explains why many individuals experience an improved mood following exercise.

While the runner's high is often associated with aerobic activity, it can also occur in other forms of exercise that elevate the heart rate and engage large muscle groups, such as dancing, hiking, or even intense yoga sessions.

Types of Exercise That Boost Endorphin Production

Not all exercises stimulate endorphins in the same way. Different forms of exercise can trigger varying levels of endorphin release depending on the type, intensity, and duration of the activity. Below, we explore several types of exercise and their endorphin-boosting potential:

- **Aerobic Exercise (Cardiovascular)**: Activities like running, cycling, swimming, and rowing are known for their ability to produce significant endorphin release. These exercises increase heart rate and promote oxygen flow to the muscles, which, in turn, triggers endorphin production to alleviate the stress caused by exertion. Aim for at least 30 minutes of moderate-to-intense aerobic exercise for optimal endorphin synthesis.

- **Strength Training**: Weightlifting and resistance training are also effective for boosting endorphin levels. While not traditionally associated with endorphin release to the same degree as aerobic exercise, strength training can stimulate significant hormonal and biochemical responses in the body. High-intensity workouts and lifting heavier weights generally yield greater endorphin benefits.

- **High-Intensity Interval Training (HIIT)**: HIIT combines short bursts of intense exercise followed by rest periods. This method has been shown to be highly effective for stimulating endorphin production. HIIT exercises, such as sprinting intervals or circuit training, can push the body to produce greater endorphin levels in response to the intense bursts of physical exertion.

- **Yoga and Mind-Body Exercises**: While yoga may not always result in high-intensity physical strain, it still plays an important role in endorphin production. Yoga, Tai Chi, and Pilates promote a sense of relaxation and stress reduction while helping to improve flexibility, strength, and balance. The controlled breathing and focus on mindfulness during these exercises stimulate the release of endorphins and help to foster emotional well-being.

- **Dancing**: Dance, especially when performed to energetic music, can significantly boost endorphin production. The rhythmic movements and full-body engagement involved in dancing increase heart rate and muscular exertion, providing a similar endorphin response to other aerobic exercises.

Training and Endorphin Optimization for Athletes

For athletes, optimizing endorphin production can enhance performance, improve recovery, and boost overall motivation. However, to achieve maximum endorphin benefits, training protocols must be tailored to the individual's needs, fitness level, and goals.

Here are some strategies for athletes to optimize endorphin production:

- **Progressive Overload**: To stimulate greater endorphin production, athletes should gradually increase the intensity or duration of their workouts. This progressive overload will push the body to produce more endorphins to cope with the increased physical demand, which in turn promotes higher levels of performance.

- **Variety in Training**: Incorporating different types of exercise, such as alternating between strength training and cardiovascular work, can help activate different physiological pathways for endorphin release. A varied training routine also helps prevent plateaus and keeps the body adapting to new challenges.

- **Active Recovery**: Active recovery, such as light jogging, walking, or swimming after intense workouts, helps maintain endorphin levels while promoting muscle repair. Rather than resting completely, light movement can keep endorphin production elevated and aid in recovery.

- **Mindfulness and Mental Focus**: For athletes, the mental aspect of exercise is just as important as the physical. Maintaining focus on positive, goal-oriented thoughts during workouts, or using visualization techniques, can stimulate endorphin release. Mindfulness practices, even during intense training, enhance the mental benefits and optimize endorphin synthesis.

- **Rest and Recovery**: While exercise is essential for boosting endorphins, overtraining can lead to burnout and negatively affect hormone levels. Proper rest, sleep, and active recovery days are essential for maintaining a balanced endorphin response, preventing depletion, and ensuring optimal long-term performance.

Conclusion: Maximizing Endorphin Production Through Exercise

Exercise is a powerful and natural way to stimulate endorphin production, providing a myriad of benefits for both physical and mental well-being. Whether through intense aerobic exercise, strength training, or more mindful practices like yoga, regular physical activity promotes endorphin release, enhancing mood, reducing pain, and improving overall resilience.

To truly harness the power of endorphins through exercise, it is important to maintain a consistent and varied workout routine that challenges both the body and the mind. By optimizing exercise routines, athletes, fitness enthusiasts, and anyone looking to improve their physical and emotional health can maximize the benefits of endorphin production, fostering a life of vitality, well-being, and balance.

In the next chapters, we will explore how other lifestyle factors such as nutrition, mental health, and mindfulness can further support endorphin production, offering additional strategies for enhancing physical and mental well-being.

Chapter 5: Mental Health and Endorphin Regulation

Endorphins, often called the body's natural "feel-good" chemicals, play a crucial role in regulating mood, emotional well-being, and mental resilience. In this chapter, we will explore the intricate relationship between endorphins and mental health, focusing on how these powerful peptides help alleviate conditions such as anxiety, depression, and stress. Additionally, we will discuss how optimizing endorphin levels can enhance mental resilience, promote emotional stability, and even provide therapeutic potential for managing mental health disorders.

The Link Between Endorphins and Mood

Endorphins are primarily known for their mood-enhancing effects. When the body produces these natural chemicals, they help create a sense of euphoria and pleasure, which can counteract negative emotional states. This mechanism is particularly beneficial in regulating mood disorders, as endorphins play a direct role in balancing emotional responses and promoting feelings of well-being.

The release of endorphins often coincides with activities that increase feelings of happiness, such as socializing, laughing, or participating in physical activities. As such, endorphins are intimately connected with the brain's **reward system**—the neural pathways that drive us to seek out pleasurable experiences and avoid discomfort. When endorphins are released, they bind to opioid receptors in the brain, triggering feelings of relaxation, joy, and overall satisfaction.

In moments of emotional or physical distress, the body turns to its internal resources—like endorphins—to restore equilibrium. By reducing the perception of pain and inducing positive feelings, endorphins help modulate the emotional intensity of stress, anxiety, and sadness, providing relief from negative feelings that can otherwise overwhelm the mind.

Endorphins in Anxiety, Depression, and Stress Relief

Endorphins and Anxiety

- **Exercise**: Regular physical activity is one of the most effective ways to naturally boost endorphin levels, providing long-term relief for anxiety. Activities like running, cycling, and yoga have been shown to reduce anxiety symptoms significantly by stimulating the production of endorphins, thereby helping to mitigate the effects of stress.
- **Mindfulness and Socialization**: Practices such as meditation, mindfulness, and social interaction can also promote endorphin release, making them valuable tools for those seeking to manage anxiety in daily life.

Endorphins and Depression

depression

neurotransmitters

serotonin

dopamine

- **Exercise**: Physical activity remains one of the most effective treatments for depression, with studies consistently showing that exercise-induced endorphin release leads to reduced depressive symptoms. Not only does exercise help to alleviate the immediate feelings of sadness, but regular activity can improve long-term mental health and help prevent relapse into depressive episodes.

- **Laughter and Social Engagement**: Social interaction and laughter are natural mood enhancers that stimulate endorphin production. These activities create positive feedback loops, whereby feelings of joy and connection enhance endorphin release, which in turn reduces symptoms of depression.

Endorphins and Stress Relief

cortisol

Endorphins act as natural antagonists to cortisol. When endorphin levels rise, cortisol levels decrease, mitigating the negative effects of stress and preventing it from overwhelming the individual. The body's ability to recover from stress and maintain emotional stability relies heavily on the balance between these two hormones.

Relaxation Techniques

How Endorphins Influence Mental Resilience

Mental resilience refers to the ability to adapt to challenges, bounce back from adversity, and maintain a positive outlook during difficult times. Endorphins play a pivotal role in building and maintaining mental resilience by influencing how the brain processes and responds to stress.

When endorphins are released, they not only alleviate the discomfort of stress and pain but also increase feelings of optimism and mental strength. This allows individuals to face challenges with greater emotional balance and reduced susceptibility to negative emotions.

- **Stress Adaptation**: Endorphins promote healthy emotional responses to stress, enabling individuals to handle future stressors more effectively. Over time, the body learns to release endorphins in response to less intense stress, reinforcing positive coping strategies and increasing resilience.

- **Self-Efficacy**: Increased endorphin levels are associated with a greater sense of **self-efficacy**—the belief in one's ability to overcome challenges and achieve goals. This feeling of empowerment bolsters mental resilience, encouraging individuals to keep moving forward in the face of adversity.

- **Emotional Regulation**: Regular endorphin boosts, whether through exercise, social engagement, or mindfulness practices, help individuals regulate their emotions more effectively. By improving emotional stability, endorphins enable better decision-making, healthier relationships, and the capacity to face life's challenges with confidence.

Therapeutic Potential of Endorphin-Based Treatments

Given the critical role that endorphins play in mental health, their therapeutic potential in the treatment of mood disorders is immense. Many conventional treatments for mental health disorders, including **antidepressants** and **anti-anxiety medications**, aim to regulate neurotransmitters such as serotonin and dopamine. However, endorphins provide a unique advantage in that they not only affect mood but also influence pain perception, stress response, and emotional resilience.

1. **Exercise as Therapy**: The use of exercise to enhance endorphin levels is increasingly being recognized as a valid therapeutic option for mental health. Cognitive-behavioral therapy (CBT), along with exercise programs, has shown promise in treating both anxiety and depression. By combining the mental health benefits of exercise with traditional therapy, individuals can experience enhanced therapeutic outcomes.

2. **Mindfulness and Meditation**: Mindfulness practices, such as meditation and deep breathing exercises, are being increasingly utilized in mental health treatment programs. These techniques enhance endorphin release and promote emotional regulation, making them valuable tools in managing mood disorders.

3. **Complementary Therapies**: Complementary therapies, such as acupuncture and massage therapy, stimulate the body's production of endorphins and can aid in the treatment of anxiety, depression, and stress. When used alongside conventional treatments, these therapies can provide additional relief and support.

Conclusion: Leveraging Endorphins for Mental Health

Endorphins are critical to maintaining mental health and emotional well-being. From alleviating the symptoms of anxiety and depression to enhancing resilience in the face of stress, endorphins help regulate mood, promote mental clarity, and foster a sense of calm. By actively engaging in practices that boost endorphin levels—such as physical exercise, mindfulness, and social engagement—individuals can optimize their mental health, reduce emotional distress, and build greater resilience. In the following chapters, we will explore other practical strategies, including nutrition and mindfulness, to further enhance endorphin production and support lifelong mental well-being.

Chapter 6: Nutrition and Endorphin Production

Nutrition plays a crucial role in regulating the production of endorphins and optimizing their availability in the body. What we eat not only provides the fuel for our cells but also impacts the biochemical processes that govern our mood, pain perception, and overall well-being. In this chapter, we will explore the relationship between food and endorphin production, focusing on the types of nutrients, amino acids, and foods that enhance endorphin synthesis. We will also examine how certain supplements can support or disrupt endorphin levels and how to create a diet that promotes mental and physical health.

Foods that Stimulate Endorphin Synthesis

Certain foods have been shown to boost the production of endorphins by providing the necessary building blocks or by triggering biochemical pathways that activate endorphin release. Here are some key food groups and their role in endorphin production:

1. **Spicy Foods**

   Spicy foods, such as chili peppers, contain **capsaicin**, a compound that triggers the body's pain receptors and stimulates endorphin production. When capsaicin is ingested, it creates a sensation of heat or burning, which is perceived as a mild form of pain. The body responds by releasing endorphins to reduce the discomfort, leading to a temporary sense of euphoria and well-being. Incorporating spicy foods into your diet can thus be a natural way to stimulate endorphin release.

2. **Dark Chocolate**

   Dark chocolate is another well-known food that promotes endorphin release. It contains compounds like **theobromine** and **phenylethylamine**, which stimulate the release of endorphins and improve mood. Studies have shown that consuming chocolate can activate the brain's reward centers, increasing feelings of pleasure and satisfaction. Dark chocolate, in particular, is rich in flavonoids, which have antioxidant properties and support brain health.

3. **Foods Rich in Omega-3 Fatty Acids**

Omega-3 fatty acids, found in foods like **salmon**, **flaxseeds**, **chia seeds**, and **walnuts**, have been shown to support brain function and promote the synthesis of neurotransmitters, including endorphins. These healthy fats help regulate inflammation in the brain and may enhance the brain's reward system, improving mood and reducing symptoms of depression.

4. **Bananas**

   Bananas are a great source of **vitamin B6**, a nutrient that plays a crucial role in the production of serotonin and endorphins. Vitamin B6 helps convert tryptophan, an amino acid found in foods like turkey and nuts, into serotonin, a neurotransmitter that has mood-lifting properties. By supporting serotonin production, bananas indirectly enhance endorphin levels and contribute to emotional well-being.

5. **Leafy Greens and Vegetables**

   Leafy greens, such as **spinach** and **kale**, are high in **folate**, a B-vitamin that is essential for maintaining healthy brain function. Folate helps the brain produce neurotransmitters like serotonin and dopamine, both of which play key roles in mood regulation and endorphin release. Additionally, leafy greens are packed with **magnesium**, which has been shown to support mood stability and promote relaxation.

6. Fermented Foods

Fermented foods like **kimchi, sauerkraut, yogurt**, and **kefir** contain beneficial probiotics that support gut health. A healthy gut microbiome is essential for the production of various neurotransmitters, including endorphins. Studies have shown that the gut-brain axis—the biochemical communication between the gut and the brain—plays a role in regulating mood and emotional well-being. Fermented foods can therefore indirectly contribute to endorphin production by promoting a balanced gut microbiome.

7. Nuts and Seeds

Nuts and seeds, such as **almonds, sunflower seeds**, and **pumpkin seeds**, are rich in **magnesium**, which supports the synthesis of neurotransmitters like serotonin and endorphins. These foods also contain **healthy fats** and **protein**, which are important for maintaining steady energy levels and ensuring the proper functioning of the nervous system. Regular consumption of nuts and seeds can help stabilize mood and enhance endorphin production.

The Role of Amino Acids in Endorphin Production

Amino acids are the building blocks of proteins, and several amino acids play a direct role in the production of endorphins and other neurotransmitters. Here are some of the most important amino acids that influence endorphin synthesis:

1. Phenylalanine

Phenylalanine is an essential amino acid found in foods like **eggs**, **meat**, **fish**, and **soy products**. It is a precursor to **tyrosine**, which is required for the production of dopamine, a neurotransmitter that works closely with endorphins in the brain's reward system. By supporting dopamine synthesis, phenylalanine can help enhance endorphin release and promote feelings of pleasure and motivation.

2. Tyrosine

Tyrosine is derived from phenylalanine and is important for the synthesis of several neurotransmitters, including **dopamine** and **noradrenaline**. While tyrosine itself does not directly contribute to endorphin production, it plays an indirect role by supporting dopamine pathways, which in turn interact with endorphins. Foods rich in tyrosine include **chicken**, **turkey**, **dairy products**, and **tofu**.

3. Tryptophan

Tryptophan is an amino acid found in foods like **turkey**, **chicken**, **pumpkin seeds**, and **oats**. It is a precursor to **serotonin**, a neurotransmitter involved in mood regulation. Tryptophan also plays an indirect role in endorphin production, as serotonin enhances the effects of endorphins in the brain. Consuming tryptophan-rich foods can help increase serotonin levels and enhance the mood-boosting effects of endorphins.

4. **Glutamine**

Glutamine is the most abundant amino acid in the body and is found in foods such as **beef**, **poultry**, **fish**, and **dairy products**. Glutamine plays a key role in supporting brain function and enhancing the body's ability to manage stress. By promoting healthy brain function, glutamine indirectly supports the synthesis of neurotransmitters like endorphins and serotonin.

Nutrient–Dense Foods for Optimal Mental Health

A nutrient-rich diet that supports brain health is essential for optimizing endorphin production and overall mental well-being. A few additional nutrients to consider:

- **Vitamin D**: Found in **fatty fish**, **eggs**, and **fortified dairy**, vitamin D supports serotonin and dopamine production, both of which enhance endorphin release.
- **B-Vitamins**: These vitamins, particularly **B6** and **B12**, help maintain healthy brain function and contribute to the production of serotonin and dopamine. Sources include **whole grains**, **leafy greens**, and **eggs**.
- **Antioxidants**: Foods rich in antioxidants, such as **berries**, **green tea**, and **dark chocolate**, help reduce oxidative stress and inflammation, both of which can disrupt neurotransmitter function. A diet rich in antioxidants supports healthy brain chemistry and promotes balanced endorphin levels.

Supplements and Their Impact on Endorphin Levels

While a well-balanced diet is the best way to support endorphin production, some people may choose to supplement their nutrition to boost endorphin levels further. Certain supplements can directly or indirectly influence endorphin synthesis:

1. **Curcumin**: The active compound in turmeric, curcumin has been shown to boost dopamine levels and reduce inflammation, which can indirectly support endorphin production.
2. **Magnesium**: As mentioned, magnesium supports the synthesis of serotonin and endorphins. Magnesium supplements can be helpful for individuals who are deficient in this essential mineral.
3. **Rhodiola Rosea**: This adaptogenic herb has been found to help balance the body's stress response and increase endorphin levels, helping to reduce anxiety and improve mood.
4. **L-Theanine**: Found in green tea, L-theanine can help promote relaxation, increase dopamine, and potentially enhance endorphin activity.

Conclusion: Creating an Endorphin-Boosting Diet

A diet rich in nutrient-dense foods, amino acids, and antioxidants is essential for supporting endorphin production and maintaining mental and physical health. By incorporating foods like spicy peppers, dark chocolate, fatty fish, and fermented foods, you can naturally stimulate the synthesis of endorphins and enhance your mood, pain management, and stress resilience. For optimal results, combine these dietary strategies with other practices, such as regular exercise, mindfulness, and proper sleep, to ensure that endorphins are consistently available for your well-being. In the next chapters, we will explore how other lifestyle factors, such as hormonal balance and stress management, can further support endorphin optimization.

Chapter 7: Hormonal Interactions: Endorphins and Other Neurotransmitters

Endorphins are part of a complex network of neurotransmitters and hormones in the body, each playing an essential role in regulating various physiological and psychological functions. Understanding how endorphins interact with other critical neurotransmitters, such as **dopamine** and **serotonin**, can provide a deeper insight into how they influence mood, behavior, and overall well-being. In this chapter, we will explore the relationships between endorphins and these other key chemicals, focusing on how hormonal balance impacts health and performance. Additionally, we will discuss strategies for optimizing the interactions between endorphins and these neurotransmitters to support peak mental and physical performance.

The Relationship Between Endorphins, Dopamine, and Serotonin

Endorphins and Dopamine

endorphins

dopamine

- **Endorphins Enhance Dopamine Release**: When endorphins are released, they activate opioid receptors that not only reduce pain and stress but also stimulate dopamine production. This synergy helps to magnify feelings of pleasure and satisfaction, creating a reinforcing loop that encourages the pursuit of rewarding activities like exercise, social engagement, and creative endeavors.

- **Motivation and Reward**: Dopamine plays a central role in motivation—essentially driving us to seek out pleasurable experiences and rewards. Endorphins enhance dopamine's effects by magnifying the pleasurable feelings associated with these activities, which encourages individuals to engage in behaviors that promote well-being, such as physical activity or engaging in meaningful social connections.

- **Exercise and Reward**: The combination of endorphins and dopamine is one reason why exercise is so effective at boosting mood. The release of endorphins during physical activity can increase dopamine production, creating the "high" that many individuals experience after a workout. This heightened dopamine activity not only provides a mood boost but also enhances motivation, making it easier to continue pursuing healthy behaviors.

Endorphins and Serotonin

- **Endorphin-Induced Serotonin Release**: Endorphins can stimulate the production of serotonin, contributing to improved mood and a sense of well-being. This interaction is particularly significant in individuals who experience conditions such as **depression, anxiety**, or **stress**, as both endorphins and serotonin help regulate emotional responses and alleviate negative mood states.

- **Mood Enhancement and Stress Relief**: When endorphins are released, they help activate the serotonin system, leading to a reduction in feelings of anxiety and stress. This mechanism helps to balance the effects of stress hormones, such as cortisol, and can contribute to improved resilience and emotional regulation.

- **Exercise and Serotonin**: Just like dopamine, serotonin is also positively influenced by physical activity. Exercise is well-known for increasing serotonin levels, which helps explain the mood-boosting effects of a workout. The combination of endorphins, dopamine, and serotonin during exercise creates an optimal environment for emotional well-being, making regular physical activity a powerful tool for managing mood disorders and maintaining a positive outlook.

Hormonal Balance and Well-Being

The balance between endorphins, dopamine, serotonin, and other hormones plays a crucial role in overall mental and physical health. These neurotransmitters do not work in isolation; instead, they form a complex network that influences everything from emotional regulation to cognitive function and physical performance. When one of these chemicals is out of balance, it can have wide-ranging effects on mood, behavior, and health.

1. **The Role of Endorphins in Hormonal Balance**

   Endorphins are not just responsible for relieving pain and improving mood—they also help maintain hormonal balance by interacting with other neurotransmitter systems. By stimulating the release of dopamine and serotonin, endorphins help regulate emotional responses and create a stable biochemical environment in the brain and body. Additionally, endorphins can moderate the levels of stress hormones, such as cortisol, which, when chronically elevated, can have negative impacts on physical and mental health.

2. **Cortisol and the Endorphin Response**

   Cortisol, often called the "stress hormone," is released during times of stress and is essential for the body's "fight or flight" response. However, prolonged high levels of cortisol due to chronic stress can lead to health issues like **insomnia**, **anxiety**, **digestive problems**, and **decreased immunity**. Endorphins help to counteract the negative effects of cortisol by promoting a relaxation response in the body. By activating opioid receptors, endorphins reduce the perception of pain and stress, thereby lowering cortisol levels and fostering emotional and physical resilience.

3. **The Importance of Balance**

An optimal balance between these neurotransmitters—endorphins, dopamine, serotonin, and cortisol—promotes emotional stability, reduces anxiety and depression, and improves cognitive function. However, imbalances in these chemicals can lead to mood disorders, stress-related illnesses, and cognitive dysfunction. For instance, a lack of endorphins or serotonin can contribute to **depression** and **anxiety**, while an excess of cortisol due to chronic stress can impair cognitive function and lead to burnout.

Achieving a balance between these hormones is essential for mental and physical well-being. Strategies to optimize this balance include physical exercise, stress management techniques, a nutrient-rich diet, and social engagement—each of which plays a role in supporting endorphin production and hormonal regulation.

Optimizing Neurotransmitter Systems for Peak Performance

To optimize the interactions between endorphins, dopamine, serotonin, and other neurotransmitters for peak performance, it is essential to engage in practices that support the production and balance of these chemicals. Below are several strategies that can help optimize the neurochemical systems in the body:

1. **Regular Exercise**: Physical activity is one of the most powerful tools for optimizing neurotransmitter balance. Exercise increases the production of endorphins, dopamine, and serotonin, creating a positive feedback loop that promotes mental clarity, emotional stability, and physical vitality.

2. **Mindfulness and Meditation**: Mindfulness practices, including meditation, deep breathing exercises, and yoga, are effective for reducing stress and promoting endorphin release. These practices also enhance serotonin production, which can improve mood and emotional regulation.

3. **Balanced Nutrition**: A well-rounded diet rich in **amino acids, vitamins**, and **minerals** supports neurotransmitter synthesis and overall brain function. Foods that are rich in tryptophan, such as turkey, nuts, and seeds, can help increase serotonin levels, while foods rich in **tyrosine** (e.g., lean meats, dairy, eggs) support dopamine production.

4. **Social Connections**: Engaging in social activities, connecting with loved ones, and participating in group exercises can stimulate endorphin and serotonin release, promoting emotional well-being. Positive social interactions help regulate mood, enhance dopamine production, and foster feelings of connection and support.

5. **Adequate Sleep**: Sleep is essential for restoring neurotransmitter balance. During sleep, the brain processes emotions and consolidates memories, which helps to optimize serotonin and dopamine systems. Getting sufficient restorative sleep is crucial for maintaining mental clarity, emotional stability, and physical health.

Conclusion: The Power of Balanced Neurotransmitter Systems

The interactions between endorphins, dopamine, serotonin, and other hormones are integral to maintaining mental and physical well-being. By optimizing the balance between these neurotransmitters, individuals can enhance their mood, emotional resilience, and cognitive function while promoting overall health. Engaging in regular physical activity, practicing mindfulness, maintaining a healthy diet, and fostering positive social relationships are key strategies for optimizing these systems and achieving peak performance. In the next chapters, we will delve deeper into additional ways to enhance endorphin production and improve well-being by focusing on sleep, stress management, and other lifestyle practices.

Chapter 8: The Endorphin Stress Response

In today's fast-paced world, stress is an inevitable part of life. Whether it's work, personal challenges, or societal pressures, stress is something we all experience to varying degrees. Fortunately, the body has an innate mechanism to cope with these stressors—**endorphins**. These natural peptides not only help mitigate pain but also play a critical role in managing stress. In this chapter, we will explore the endorphin stress response, how endorphins help the body cope with stress, the neurobiological mechanisms behind stress and endorphins, and strategies for activating endorphins during stressful situations.

How Endorphins Help the Body Cope with Stress

Stress, whether acute or chronic, triggers a cascade of physiological responses designed to help the body adapt to or manage the perceived threat. When the body is under stress, it produces **stress hormones** like **cortisol** and **adrenaline** (epinephrine), which prepare the body for the "fight or flight" response. While these hormones are essential in the short term for survival, prolonged stress can have detrimental effects on health, leading to conditions such as hypertension, weakened immune function, and mental health disorders like anxiety and depression.

Endorphins act as the body's natural countermeasure to the stress response. As part of the **hypothalamic-pituitary-adrenal (HPA)** axis, which governs the body's reaction to stress, endorphins help regulate and balance the production of cortisol and other stress-related hormones. Their key role is to **mitigate the physical and emotional impact of stress** by promoting relaxation, reducing anxiety, and enhancing feelings of well-being.

- **Pain and Stress Management**: When under stress, the body often experiences tension, whether it's muscle tightness or headaches. Endorphins act as natural analgesics, helping to reduce these physical manifestations of stress by binding to opioid receptors in the brain and nervous system. This action not only alleviates pain but also soothes the nervous system, providing an immediate sense of relief.

- **Reducing Cortisol**: Chronic stress leads to sustained high levels of cortisol, which can negatively affect health. Endorphins help counteract the effects of cortisol by moderating the body's stress response. They work synergistically with the parasympathetic nervous system, which is responsible for calming the body after stress and restoring balance.

- **Emotional Resilience**: Endorphins contribute to emotional resilience by enhancing the brain's ability to cope with stressors. They reduce the emotional intensity of stress, enabling individuals to maintain a positive outlook and cope more effectively with difficult situations.

The Neurobiological Mechanism of Stress and Endorphins

The body's stress response is controlled by a network of hormones, neurotransmitters, and the autonomic nervous system. Endorphins play an integral role in this system by influencing both the **central nervous system (CNS)** and **peripheral nervous system (PNS)**.

1. **Activation of the Hypothalamic-Pituitary-Adrenal (HPA) Axis**:

   The HPA axis is the body's primary stress response system. When the brain perceives stress, it signals the hypothalamus to release corticotropin-releasing hormone (CRH), which activates the pituitary gland to release adrenocorticotropic hormone (ACTH). ACTH stimulates the adrenal glands to release cortisol, the primary stress hormone. This cascade of reactions helps the body prepare for action.

   In response to stress, endorphins are also released from the **pituitary gland**, which plays a central role in regulating the body's reaction to pain and stress. When endorphins are released, they bind to opioid receptors in the brain, leading to a reduction in pain perception, anxiety, and stress levels. This interplay between endorphins and cortisol helps moderate the body's stress response, ensuring that cortisol levels do not remain elevated for prolonged periods.

2. **Opioid Receptors and the Stress Response**:

   Endorphins exert their effects by binding to specific **opioid receptors** in the brain, primarily **mu** and **delta** receptors. These receptors are involved in pain relief, mood regulation, and stress adaptation. By activating these receptors, endorphins reduce the perception of pain and promote relaxation, both of which are essential for managing stress.

- **Mu receptors** are primarily responsible for the analgesic (pain-relieving) effects of endorphins, as well as for the euphoric and calming effects.
- **Delta receptors** are thought to play a role in mood regulation, helping to stabilize emotions during stressful events.

Dopamine, Serotonin, and Endorphins

dopamine

serotonin

Strategies for Activating Endorphins During Stressful Situations

There are several ways to activate endorphin production during stressful situations. Whether you're experiencing a high-pressure work deadline, dealing with personal challenges, or facing a physically stressful event, there are practical steps you can take to increase your endorphin levels and reduce the impact of stress on your mind and body.

Physical Activity

Breathing Exercises

diaphragmatic breathing

4-7-8 technique

2. **Socializing and Laughter**:

   Positive social interactions and laughter are powerful activators of endorphins. Spending time with loved ones, engaging in enjoyable conversations, or even laughing out loud can trigger the release of these feel-good chemicals. In fact, laughter has been shown to reduce cortisol levels and increase endorphin production, leading to improved mood and reduced stress.

3. **Mindfulness and Meditation**:

   Mindfulness practices, including meditation, yoga, and tai chi, are excellent tools for managing stress and boosting endorphins. These techniques involve focusing on the present moment and calming the mind, which not only reduces stress but also promotes the release of endorphins. Studies have shown that mindfulness practices increase serotonin and endorphin levels, contributing to emotional balance and improved stress management.

4. **Music and Dance**:

   Listening to music or engaging in dance movements can also stimulate endorphin release. Upbeat, rhythmic music, in particular, has been shown to reduce cortisol levels and elevate mood. Dancing, as a form of physical exercise combined with music, is especially effective in promoting endorphin release. It enhances the body's ability to cope with stress while providing a fun and creative outlet for expression.

5. **Aromatherapy**:

Aromatherapy, particularly the use of essential oils like **lavender**, **bergamot**, and **chamomile**, can reduce stress and promote relaxation. These scents have been shown to stimulate the release of endorphins and other calming chemicals, such as serotonin, providing immediate relief from stress.

6. **Visualization and Positive Thinking**:

Mental techniques, such as visualization, can activate the endorphin response by triggering feelings of calm, peace, and joy. Imagining a peaceful, relaxing scene, or visualizing success in a stressful situation, can reduce anxiety and activate the body's natural relaxation response. Positive thinking and focusing on gratitude also promote the release of endorphins by shifting the brain's focus from stress to positive experiences.

Conclusion: Endorphins as the Body's Natural Stress Reliever

Endorphins are integral to managing stress and promoting emotional well-being. As natural painkillers and mood enhancers, endorphins reduce the physiological and emotional impact of stress, creating a sense of balance and relaxation. By activating endorphins through exercise, socializing, mindfulness, and other techniques, individuals can enhance their ability to cope with life's challenges and foster resilience in the face of adversity. In the next chapters, we will continue to explore how lifestyle factors, such as sleep and environmental factors, further support endorphin production and overall well-being.

Chapter 9: Sleep and Endorphin Levels

Sleep is one of the most crucial aspects of health, not only for physical restoration but also for mental well-being. It is during sleep that the body undergoes essential processes of repair, memory consolidation, and hormonal regulation. Among these, **endorphins** play a significant role in ensuring that the sleep-wake cycle functions optimally. In this chapter, we will explore how sleep influences endorphin production, the specific role of **REM sleep** in endorphin synthesis, and strategies to optimize sleep for better endorphin availability and overall health.

How Sleep Affects Endorphin Production

Sleep and endorphins are deeply interconnected. When we sleep, particularly during deep and REM sleep, the body's natural processes are in full swing, including the production of hormones and neurotransmitters. Endorphins, which are primarily known for their roles in pain relief and mood regulation, also contribute to better sleep quality and overall recovery from physical and emotional stress.

1. **Sleep-Induced Endorphin Release**:

   Throughout the sleep cycle, the brain and body experience various states of activity. Endorphins are released at key points to support relaxation and recovery. For instance, during the **deep sleep stages** (also known as **slow-wave sleep**), endorphins help to reduce stress hormones like **cortisol**, which naturally decline during sleep. The presence of endorphins during sleep promotes a deeper, more restorative sleep cycle by counteracting stress and supporting the body's recovery processes.

2. **Endorphins and Pain Reduction During Sleep**:

   One of the benefits of sleep is the reduction of perceived pain, which is partly mediated by endorphin production. While the body rests, the natural release of endorphins helps to reduce physical discomfort and enhances the body's ability to repair itself. This is especially beneficial for those experiencing chronic pain or recovering from injuries, as endorphins assist in reducing pain perception and promoting healing.

3. **Regulation of Emotional Stress**:

   Endorphins also help to regulate emotional stress and mood, which is why sleep is so important for mental well-being. When we don't get enough sleep, stress hormones like cortisol can build up, leading to increased anxiety, irritability, and fatigue. Adequate sleep, on the other hand, allows endorphins to act as natural mood enhancers, helping to promote feelings of relaxation and happiness when we wake up. This contributes to emotional resilience and supports overall mental health.

The Role of REM Sleep in Endorphin Synthesis

Sleep occurs in cycles, with each cycle lasting about 90 minutes, and each cycle includes various stages of light sleep, deep sleep, and **Rapid Eye Movement (REM) sleep**. REM sleep, the phase in which dreams occur, is particularly important for endorphin synthesis and emotional well-being.

Endorphin Release During REM Sleep

Memory and Emotional Health

2. **The Neurobiological Mechanisms of REM Sleep and Endorphins**:

   The interaction between **neurotransmitters** and **hormones** during REM sleep is complex. During this phase, endorphins are released in response to neural activity and help to regulate the emotional and physical responses of the body. In particular, endorphins interact with **dopamine** and **serotonin**, both of which are vital for maintaining mood balance and mental clarity. This neurobiological mechanism supports the integration of positive emotions and mitigates the impact of negative emotional states, enhancing overall psychological resilience.

3. **Restoring Mental Health Through REM Sleep**:

   Given the important role REM sleep plays in processing emotional experiences, it is during this stage that endorphins are most effective at relieving emotional stress and promoting emotional recovery. People who experience insufficient REM sleep often report higher levels of anxiety, depression, and irritability. By ensuring that you get enough quality REM sleep, you enable endorphins to perform their vital function in resetting emotional states and supporting mental well-being.

Sleep Optimization Techniques to Support Endorphin Production

Since sleep is so vital to endorphin production and overall well-being, optimizing your sleep habits is essential for maximizing the benefits of endorphins. Below are several strategies to support healthy sleep and ensure endorphins are available in adequate quantities to help reduce stress, improve mood, and enhance recovery:

1. **Maintain a Consistent Sleep Schedule**

   One of the most effective ways to improve sleep quality is by sticking to a consistent sleep schedule. Going to bed and waking up at the same time every day helps regulate the body's **circadian rhythm**, ensuring that you get adequate amounts of both deep and REM sleep. A consistent sleep schedule helps synchronize your internal clock with natural sleep-wake cycles, optimizing hormone production, including endorphins.

2. **Create a Sleep-Conducive Environment**

   Your sleep environment plays a significant role in the quality of sleep you get. To enhance endorphin production during sleep, make sure your bedroom is cool, dark, and quiet. Consider using blackout curtains to block out light and noise machines or earplugs to minimize disturbances. A comfortable mattress and pillow that support your sleeping posture will also contribute to restful, uninterrupted sleep.

3. **Limit Stimulants and Distractions**

   Avoid consuming **caffeine**, **nicotine**, and **heavy meals** at least four hours before bedtime, as these can disrupt sleep by stimulating the nervous system. Additionally, reduce your exposure to **blue light** emitted by screens (phones, computers, televisions) at least 30 minutes to an hour before bedtime, as blue light interferes with the production of **melatonin**, the hormone that regulates sleep.

4. **Practice Relaxation Techniques**

   Engaging in relaxation techniques, such as **deep breathing, progressive muscle relaxation**, or a calming pre-sleep routine, can help signal to your body that it is time to wind down. Techniques like **meditation** or **gentle stretching** can also help reduce stress and anxiety, setting the stage for better sleep. By lowering cortisol levels, these practices allow your body to enter the sleep state more easily, optimizing endorphin production and sleep quality.

5. **Exercise Regularly**

   Regular physical activity during the day can improve the quality of your sleep, especially if you engage in moderate-to-intense exercise. Exercise helps regulate sleep patterns by reducing stress, promoting endorphin release, and preparing the body for restful sleep. However, it is important to avoid intense exercise too close to bedtime, as it can increase adrenaline levels and potentially interfere with your ability to fall asleep.

6. **Consider Sleep Supplements**

   If you have difficulty achieving restful sleep, certain supplements can support the process. **Melatonin** is a well-known supplement that can help regulate sleep cycles, while **magnesium** and **L-theanine** can help relax the body and calm the

Conclusion: Enhancing Endorphin Production Through Sleep
   nervous system. These supplements can enhance the natural sleep process, supporting endorphin production during the night.

Sleep is not only a time for rest; it is a vital process that supports the production and availability of endorphins, promoting mental and physical health. By ensuring you get sufficient, high-quality sleep, especially deep and REM sleep, you enable your body to leverage the full benefits of endorphins in reducing stress, managing emotions, and enhancing recovery. Optimizing sleep through consistent routines, a relaxing environment, and mindful practices can significantly improve endorphin production and overall well-being. As we continue to explore strategies for enhancing endorphins in the next chapters, remember that sleep serves as the foundation for many of these processes, making it one of the most powerful tools for improving your health and happiness.

Chapter 10: The Science of Pleasure: Endorphins and Reward Systems

The experience of pleasure is an essential part of human existence, driving our behavior and motivating us to engage in activities that promote survival, well-being, and emotional satisfaction. At the core of this process are the **endorphins**, powerful chemicals that influence our sense of pleasure, reward, and motivation. In this chapter, we will explore the role of endorphins in the brain's reward systems, how they influence motivation, achievement, and behavior, and the ways in which we can leverage endorphin production to maximize our pleasure and well-being.

Understanding the Reward System in the Brain

The **reward system** is a complex network of brain structures that are activated by pleasurable stimuli. This system includes key regions such as the **ventral tegmental area (VTA)**, **nucleus accumbens**, and the **prefrontal cortex**, all of which work together to process rewarding experiences and encourage behaviors that are beneficial for survival.

Endorphins are critical players in this system. When pleasurable experiences occur—whether through physical activity, social interactions, or satisfying food—endorphins are released as part of the brain's response. They act as **neurotransmitters**, sending signals that promote positive emotions and reinforce the desire to repeat those behaviors. This system encourages us to pursue activities that bring us pleasure, whether they involve physical sensations (like eating or exercise) or emotional rewards (such as social connection or accomplishment).

The primary function of the reward system, and by extension endorphins, is to create a **reinforcement loop**—a cycle of positive feedback that drives us to engage in behaviors that promote survival and happiness. This loop not only enhances the immediate experience of pleasure but also encourages us to repeat the activity, fostering long-term well-being.

The Role of Endorphins in Pleasure and Reward

Endorphins are sometimes referred to as "natural painkillers" because they help to mitigate pain and discomfort. However, their role extends beyond simply reducing physical pain. Endorphins are also central to the brain's reward system, enhancing feelings of pleasure and satisfaction. Here's how they contribute to the process of pleasure and reward:

Enhancing the Pleasure Experience

opioid receptors

- **Exercise and "Runner's High"**: One of the most well-known examples of endorphins in the reward system is the "**runner's high**." This phenomenon occurs after prolonged or intense physical activity, where endorphins flood the brain, creating a sense of euphoria. This helps athletes push through physical pain and motivates them to engage in exercise regularly.

- **Food and Pleasure**: Foods high in sugar, fat, or even spicy ingredients (like chili peppers) can also stimulate endorphin release, contributing to the pleasure we feel when eating. The brain recognizes the reward from satisfying hunger and ensures we seek these experiences again in the future.

Social Bonding and Connection

Oxytocin and Endorphins

oxytocin

Motivation and Achievement

Goal Setting and Achievement

Pleasure and Risk-Taking Behavior

Adrenaline and Endorphins

adrenaline

How Endorphins Impact Motivation and Achievement

The role of endorphins in motivation is essential not only for the immediate satisfaction they provide but also for the long-term benefits of goal achievement. By driving us to pursue rewarding activities, endorphins shape our behavior in ways that promote personal success, happiness, and growth.

1. **Reinforcing Positive Behaviors**

   Endorphins act as natural reinforcers of behaviors that lead to pleasure, whether they involve physical, emotional, or social rewards. When we complete a task, receive social validation, or experience a positive outcome, endorphins are released as part of the brain's reward system. This creates a cycle where pleasurable experiences encourage us to pursue similar activities in the future, leading to further personal and professional growth.

2. **Fostering Long-Term Success**

   Long-term achievement and personal success are built on a foundation of consistent effort, which is often driven by motivation. Endorphins provide the emotional boost needed to sustain effort over time. They create feelings of fulfillment and joy when pursuing long-term goals, which can help us overcome setbacks, stay focused, and keep striving for success. In this way, endorphins act as an emotional and physiological support system, enabling us to stay on course even when faced with challenges.

3. **The Link Between Endorphins and Flow States**

Endorphins also play a key role in helping individuals enter a state of **flow**, a mental state characterized by deep immersion, focus, and enjoyment during an activity. Whether during athletic performance, creative work, or problem-solving, flow is often accompanied by a surge of endorphins that enhances performance and leads to a sense of achievement and satisfaction. By optimizing endorphin production, individuals can more easily access this state of peak performance.

Leveraging Endorphins for Optimal Pleasure and Well-Being

Understanding the role of endorphins in pleasure and reward allows us to leverage their effects to enhance our quality of life. Here are some practical strategies to optimize endorphin production for a happier, more motivated, and fulfilling life:

1. **Engage in Regular Exercise**: Physical activity is one of the most effective ways to boost endorphins and increase pleasure. Aim for a variety of exercises, from cardio to strength training, to keep your workouts enjoyable and stimulating.

2. **Cultivate Positive Social Connections**: Spend time with friends and loved ones, engage in meaningful conversations, and participate in social activities that promote bonding and laughter. These interactions can significantly enhance endorphin levels and improve emotional health.

3. **Pursue Meaningful Goals**: Setting and achieving goals, whether personal, professional, or creative, triggers endorphin release and boosts motivation. Celebrate small victories along the way to keep your motivation high.

4. **Indulge in Pleasurable Activities**: Whether it's enjoying a favorite hobby, eating a delicious meal, or simply relaxing in a hot bath, indulge in activities that bring you joy and satisfaction. These moments of pleasure activate the reward system and improve overall well-being.

Conclusion

Endorphins are powerful chemicals that enhance our experience of pleasure, reward, and motivation. By understanding their role in the brain's reward system, we can harness the power of endorphins to drive positive behaviors, increase productivity, and improve emotional and physical well-being. Whether through exercise, social interactions, or the pursuit of personal goals, endorphins are essential to creating a fulfilling and motivated life. In the next chapters, we will continue exploring additional strategies to maximize endorphin production and enhance overall health and happiness.

Chapter 11: Environmental Factors Affecting Endorphin Production

The environment in which we live and work plays a significant role in regulating our physical and emotional well-being. Factors like sunlight, temperature, noise, and exposure to nature can all influence the synthesis and availability of endorphins in the body. In this chapter, we will explore how various environmental factors affect endorphin production, how nature influences our mental and physical states, and practical steps you can take to create an **endorphin-friendly living space** that enhances overall well-being.

Sunlight and Endorphin Production

One of the most powerful environmental influences on endorphin production is **sunlight**. Natural light is crucial not only for regulating our circadian rhythm but also for promoting the production of several key neurotransmitters, including **serotonin**, **dopamine**, and, of course, **endorphins**.

1. **The Role of Sunlight in Mood Regulation**

   Sunlight helps stimulate the production of **serotonin**, which is often referred to as the "happiness hormone." Higher serotonin levels have a positive impact on mood, but sunlight also plays a key role in the production of endorphins, the body's natural painkillers and mood enhancers. Studies have shown that exposure to natural light triggers the brain's reward systems, enhancing endorphin levels and promoting feelings of happiness and relaxation.

2. **Vitamin D and Endorphin Synthesis**

   Exposure to sunlight promotes the production of **vitamin D** in the skin, which is essential for various bodily functions, including immune health and mood regulation. Research suggests that vitamin D deficiency is associated with an increased risk of depression and other mood disorders, as it may impact the availability of endorphins and other mood-regulating neurotransmitters. Spending time outdoors in natural sunlight can help increase vitamin D levels and support endorphin synthesis, especially during the brighter months when sunlight is abundant.

3. **Circadian Rhythm and Endorphin Balance**

   The body's natural circadian rhythm, which regulates the sleep-wake cycle, is directly influenced by sunlight. A well-regulated circadian rhythm promotes restorative sleep, reduces stress, and enhances the release of endorphins. Ensuring adequate exposure to natural light during the day, particularly in the morning, helps synchronize the body's internal clock, improving sleep quality and the balance of endorphins.

Temperature and Endorphin Levels

Temperature also plays a vital role in the body's response to stress and its ability to regulate mood through endorphins. Extreme temperatures—both hot and cold—trigger the body to release endorphins as part of the stress response, helping to mitigate physical discomfort and emotional stress.

1. **Cold Exposure and Endorphin Production**

   Cold exposure, such as taking a cold shower, swimming in cold water, or using ice baths, can stimulate the release of endorphins. This is because the body perceives cold exposure as a mild stressor, which activates the sympathetic nervous system and triggers endorphin release as part of the "fight or flight" response. This sudden shock to the system can lead to a surge of euphoria, commonly referred to as the **"cold rush."**

   Cold exposure has become increasingly popular as a recovery tool among athletes and wellness enthusiasts due to its ability to enhance circulation, reduce inflammation, and release endorphins. Studies have shown that regular exposure to cold temperatures can help increase the body's ability to manage stress and boost overall resilience.

2. **Warmth and Relaxation**

On the opposite end of the spectrum, warmth can promote relaxation and support endorphin production by activating the **parasympathetic nervous system**, which is responsible for calming the body. **Warm baths**, saunas, or simply spending time in a warm environment can help reduce stress, ease muscle tension, and promote the release of endorphins that help us relax and feel at ease.

Massage therapy is another form of warmth-related therapy that increases blood flow, reduces stress, and stimulates endorphin production. The gentle pressure applied during a massage can reduce tension and provide a soothing, pleasure-enhancing effect.

Nature and Endorphin Production

Spending time in nature is one of the most effective ways to boost mental and physical health. Research has consistently shown that **green spaces** and **natural environments** have a profound impact on our mood, reducing anxiety and promoting feelings of calm and happiness. Nature not only provides visual and sensory stimuli that improve well-being but also helps activate endorphins, enhancing both physical and emotional resilience.

1. **Forest Bathing and Natural Environments**

   Known as **shinrin-yoku** in Japan, "forest bathing" involves immersing oneself in the natural environment of a forest, paying attention to the sights, sounds, and smells around you. Research has demonstrated that this practice has a significant impact on mental health, reducing cortisol levels and increasing the production of endorphins. The act of walking through a forest or spending time in natural environments lowers stress and promotes feelings of relaxation and contentment.

2. **The Healing Power of Green Spaces**

   Studies have shown that simply being surrounded by plants and greenery can enhance endorphin production. The calming effect of nature has a positive impact on brain function, boosting mood, reducing stress, and fostering social connections. Urban parks, gardens, and other green spaces are valuable tools for improving mental health and promoting the release of endorphins, especially in highly stressful urban environments.

Air Quality and Endorphins

Creating an Endorphin-Friendly Living Space

The environment in which you live can either support or hinder your ability to produce endorphins effectively. A space that is both relaxing and stimulating can promote a positive mental state and support overall health. Here are some tips for creating an **endorphin-friendly living space**:

1. **Maximize Natural Light**

   Ensure your living space has plenty of access to natural light, especially in the morning. Open curtains during the day to allow sunlight to flood the room. If you work indoors, try to sit near windows or spend time outdoors during breaks. Exposure to natural light in the morning helps regulate circadian rhythms and supports healthy endorphin levels.

2. **Incorporate Greenery**

   Bring nature into your home by adding plants and flowers. Studies have shown that even a small amount of greenery in a room can improve mood and reduce stress. Houseplants like **peace lilies**, **snake plants**, and **pothos** not only improve air quality but also have a calming, restorative effect on the mind and body.

3. **Create a Relaxing Space**

   Set up a designated area in your home where you can relax and recharge. This could be a cozy corner with soft lighting, comfortable seating, and calming colors. Consider adding elements such as candles, incense, or essential oils, which can help create a serene environment that promotes relaxation and endorphin release.

4. **Optimize Temperature**

   Keep your home at a comfortable temperature, making it a space conducive to both relaxation and activity. Experiment with different environments—whether it's a cozy, warm atmosphere for relaxation or a cooler space for invigorating activity —and observe how your body responds to each.

5. **Engage in Outdoor Activities**

   Make time to regularly engage in outdoor activities, whether it's walking, hiking, or simply sitting in a park. The combination of fresh air, physical movement, and exposure to nature is a powerful catalyst for endorphin production.

Conclusion: Harnessing the Power of Your Environment

The environment you live in and the natural elements you are exposed to play a pivotal role in regulating your endorphin levels and overall well-being. By optimizing your exposure to sunlight, creating a connection with nature, and adjusting temperature conditions to promote relaxation or invigoration, you can significantly enhance endorphin production. Creating an environment that supports endorphin release empowers you to improve your physical health, emotional resilience, and overall happiness. In the next chapters, we will continue exploring additional strategies and techniques for mastering endorphins and leveraging them to create lasting well-being.

Chapter 13: The Role of Meditation and Mindfulness

Meditation and mindfulness practices have been shown to have a profound impact on mental and physical health, and one of the key benefits is their ability to boost the production and availability of **endorphins**. These practices involve intentional focus and awareness of the present moment, and they activate various neurobiological pathways that promote relaxation, stress relief, and emotional balance. In this chapter, we will explore how meditation and mindfulness stimulate endorphin production, the neurobiological mechanisms behind these effects, and practical techniques to integrate these practices into daily life for optimal well-being.

How Meditation Stimulates Endorphin Production

Meditation involves focusing the mind, often on a specific object, thought, or sensation, while calmly accepting whatever thoughts arise. This focus helps to quiet the mental chatter that can cause stress and anxiety. Over time, regular meditation practice trains the brain to respond more calmly to stress, reducing the need for the fight-or-flight response, which is mediated by stress hormones like **cortisol** and **adrenaline**.

1. **Relaxation Response and Endorphins**

   Meditation triggers what is known as the **relaxation response**, a physiological state that opposes the body's stress response. When activated, this response helps lower blood pressure, reduce heart rate, and decrease muscle tension. More importantly, the relaxation response also encourages the release of **endorphins** and **serotonin**, both of which promote feelings of calm, happiness, and satisfaction. Research has shown that even short periods of meditation can significantly increase endorphin levels, leading to a positive shift in mood and well-being.

2. **Mindfulness and Emotional Balance**

   Mindfulness is a specific type of meditation that involves paying attention to the present moment with full awareness, without judgment. This practice can lead to an increase in endorphins by promoting emotional regulation and reducing the physiological effects of stress. As mindfulness reduces negative emotional states like anxiety and anger, it activates endorphin release, helping to induce a calm, positive emotional state. Studies have found that individuals practicing mindfulness experience a greater sense of well-being and are less likely to experience depression, anxiety, or chronic stress.

The Neurobiological Effects of Mindfulness on Endorphins

Mindfulness and meditation are not just psychological exercises—they have real, measurable effects on the brain and body. These practices have been shown to change brain activity, hormone levels, and the way the body responds to stress, all of which influence endorphin production.

1. **The Brain's Reward Pathways**

   Meditation and mindfulness engage the brain's **reward system**, which is responsible for regulating feelings of pleasure and satisfaction. As endorphins are released during mindfulness practices, they interact with key regions of the brain such as the **prefrontal cortex, amygdala**, and **ventral striatum**, which are involved in emotional regulation and pleasure. This interaction helps individuals feel more at peace and improves their ability to handle stress.

   The **prefrontal cortex** plays a key role in controlling thoughts and emotions, while the **amygdala** is responsible for processing emotional responses, including fear and anxiety. Mindfulness practice reduces the activity of the amygdala, which often overreacts to stressors, while increasing activity in the prefrontal cortex, which helps regulate emotional responses. The result is a greater sense of emotional balance and a boost in endorphin levels.

2. **Autonomic Nervous System Regulation**

Meditation also activates the **parasympathetic nervous system**, which is responsible for "rest and digest" functions. This system opposes the sympathetic nervous system, which governs the "fight or flight" response. By activating the parasympathetic system, mindfulness helps to lower cortisol and adrenaline levels, while simultaneously increasing endorphin release. This balance contributes to a calm, peaceful state of mind and body, reducing both physical and emotional stress.

Mindfulness Techniques for Emotional and Physical Health

Integrating mindfulness and meditation into daily life can have significant benefits for both physical and emotional health, including enhanced endorphin production. Below are some practical mindfulness techniques that can help you boost your endorphin levels and improve overall well-being:

Mindful Breathing

mindful breathing

How to practice mindful breathing

- Find a comfortable position, either sitting or lying down.

- Close your eyes and take a few deep breaths to settle into the practice.

- Focus your attention on the sensation of the air entering and leaving your body.

- If your mind starts to wander, gently bring your focus back to your breath.

Body Scan Meditation

How to practice a body scan

- Lie down in a comfortable position and close your eyes.

- Begin by focusing on your feet and slowly move your attention upwards, scanning each part of your body.

- Pay attention to any sensations of tension, tightness, or discomfort.

- Take deep breaths and consciously relax each part of your body as you move upwards.

Loving-Kindness Meditation (Metta)

Metta

How to practice loving-kindness meditation

- Sit in a comfortable position and close your eyes.

- Begin by directing kind and loving thoughts toward yourself, repeating phrases such as, "May I be happy, may I be healthy, may I be at peace."

- Then, gradually extend these thoughts toward others, including loved ones, acquaintances, and even those you may have difficulty with.

- Focus on the warmth and love that these thoughts generate.

Walking Meditation

How to practice walking meditation

- Find a quiet, safe place to walk, either indoors or outdoors.

- Walk at a slow, deliberate pace, focusing on the sensation of your feet touching the ground.

- Pay attention to your breath and the environment around you, taking in the sights, sounds, and smells.

- If your mind begins to wander, gently bring your focus back to the act of walking and breathing.

The Neurobiological Effects of Meditation and Mindfulness

Regular meditation and mindfulness practices not only increase endorphin levels but also enhance brain function and emotional resilience. These practices have been shown to change brain structure and activity, improving areas involved in emotional regulation, decision-making, and self-awareness.

1. **Increased Gray Matter Density**

   Research has demonstrated that long-term meditation leads to increased **gray matter density** in the brain, particularly in regions responsible for emotional regulation, memory, and self-control, such as the **prefrontal cortex** and **hippocampus**. These changes support better emotional resilience and improved stress management, allowing for more effective endorphin regulation.

2. **Improved Emotional Regulation**

   Mindfulness practices help individuals develop better emotional regulation by increasing activity in the **prefrontal cortex**, which is involved in decision-making and impulse control, while reducing activity in the **amygdala**, the brain's "fear center." This neurobiological shift allows individuals to experience less anxiety, react to stress more calmly, and maintain emotional balance, all of which contribute to higher endorphin levels.

Conclusion: Enhancing Well-Being Through Meditation and Mindfulness

Meditation and mindfulness practices provide a powerful and natural way to enhance endorphin production, improve emotional health, and manage stress. By incorporating mindfulness techniques like deep breathing, body scanning, and loving-kindness meditation into your daily routine, you can create a state of mental and physical well-being that promotes relaxation, emotional resilience, and long-term health. The neurobiological effects of these practices further support endorphin regulation, contributing to a happier, more balanced life. In the next chapters, we will explore additional ways to optimize endorphins and improve overall well-being through various lifestyle practices.

Chapter 14: Endorphins in Aging and Longevity

Aging is an inevitable part of life, but the process of aging doesn't have to be synonymous with deterioration in health, vitality, or well-being. The presence of **endorphins** plays a significant role in the aging process, influencing everything from physical health to emotional resilience. In this chapter, we will explore how endorphins impact the aging process, how they help maintain vitality and cognitive function, and strategies for preserving or enhancing endorphin levels as you age.

The Impact of Endorphins on Aging Processes

Endorphins are known for their role in pain management, mood regulation, and stress relief, but they also have important effects on the body's aging process. As we age, the body naturally experiences a decline in many physiological functions, including a decrease in hormone levels and a reduction in the body's ability to produce endorphins. However, maintaining healthy endorphin levels can mitigate some of the negative aspects of aging, improve quality of life, and promote longevity.

Neuroprotective Effects of Endorphins

Endorphins and Neuroplasticity

hippocampus

Mood Regulation and Emotional Resilience

Social Connection and Endorphins

Pain Management and Physical Function

Exercise and Endorphins for Pain Relief

Maintaining Endorphin Levels as You Age

As we age, the body's natural production of endorphins tends to decline. However, there are various strategies and lifestyle adjustments that can help maintain or even enhance endorphin production throughout the aging process. Below are some practical tips to optimize endorphin levels and promote healthy aging:

Regular Physical Activity

- **Aerobic Exercise**: Activities like brisk walking, cycling, or swimming can help maintain cardiovascular health, improve endurance, and stimulate endorphin release.
- **Strength Training**: Incorporating resistance exercises into your routine can help maintain muscle mass, improve bone density, and enhance overall physical strength, all while boosting endorphin levels.
- **Stretching and Yoga**: Stretching exercises and yoga, which focus on flexibility, balance, and relaxation, are particularly beneficial for aging bodies. These activities reduce physical discomfort, increase mobility, and support endorphin production.

Social Engagement

- **Volunteer Work**: Engaging in community service or volunteering can provide a sense of purpose, reduce loneliness, and trigger endorphin release.
- **Group Activities**: Participating in group activities, whether through religious organizations, hobbies, or support groups, provides a sense of belonging and contributes to endorphin production.

Mindfulness and Meditation

- **Breathing Exercises**: Focusing on deep, slow breaths can help activate the parasympathetic nervous system, reduce stress, and boost endorphin levels.
- **Mindfulness Meditation**: By focusing on the present moment and letting go of negative thoughts, mindfulness meditation helps activate the reward system in the brain, promoting the release of endorphins and enhancing overall emotional health.

Nutrition and Supplementation

- **Foods Rich in Omega-3 Fatty Acids**: Omega-3s, found in foods like fatty fish, flax seeds, and walnuts, have been shown to support cognitive function and reduce the risk of age-related mental decline.

- **Dark Chocolate**: Dark chocolate, rich in antioxidants, is known to promote the release of endorphins and improve mood.

- **Turmeric**: The active compound in turmeric, **curcumin**, has anti-inflammatory properties that help reduce pain and promote overall health, while also stimulating the production of endorphins.

5. Additionally, certain supplements such as **vitamin D**, **magnesium**, and **B vitamins** have been shown to support endorphin production and brain health, particularly in older adults who may be deficient in these nutrients.

6. **Adequate Sleep**

   Quality sleep is essential for endorphin production, as it helps the body restore and rejuvenate. During sleep, endorphin levels rise, supporting both physical and emotional well-being. Ensuring that you get adequate sleep each night is crucial for maintaining endorphin levels and overall health.

Sleep Hygiene

Strategies for Promoting Healthy Aging with Endorphins

Incorporating endorphin-boosting activities into your lifestyle is not only beneficial for aging well but also for promoting longevity. Regular physical activity, social engagement, mental stimulation, and self-care are all important aspects of healthy aging. By actively focusing on endorphin optimization, you can maintain a high quality of life, protect your brain health, and reduce the impact of aging on your physical and emotional well-being.

1. **Consistency is Key**: Consistently practicing endorphin-boosting activities such as exercise, socializing, mindfulness, and healthy eating can make a significant impact on the aging process. Aim to integrate these practices into your daily routine to ensure sustained benefits.

2. **Stay Adaptable**: As you age, your needs and abilities may change. Adapt your exercise and social activities to suit your current physical and emotional state while continuing to prioritize endorphin optimization.

3. **Prioritize Emotional Health**: Managing stress and maintaining emotional resilience is just as important as physical health. Engage in activities that promote relaxation, emotional balance, and social connection to support long-term well-being.

Conclusion

Endorphins are crucial for promoting healthy aging, supporting both physical health and emotional well-being. By understanding how endorphins contribute to aging processes and actively working to optimize their production, you can improve your quality of life, maintain cognitive function, and reduce the effects of aging on your body and mind. Regular physical activity, social connection, mindful practices, and a healthy diet all play essential roles in maximizing endorphin levels, ultimately allowing you to enjoy a longer, healthier, and more fulfilling life.

Chapter 15: Genetics and Endorphin Response

Genetics play a crucial role in determining how our bodies produce, respond to, and utilize endorphins. Each individual has a unique genetic blueprint that influences the number and sensitivity of endorphin receptors, the efficiency of their synthesis pathways, and the overall effectiveness of endorphins in regulating pain, mood, and stress. This chapter explores how genetic variations affect endorphin production, the mechanisms underlying these variations, and personalized strategies for optimizing endorphin levels based on genetic predispositions.

Genetic Variations in Endorphin Receptors

The interaction between endorphins and their receptors is a critical component of their effectiveness. Endorphins bind to **opioid receptors** in the brain and nervous system, which are responsible for mediating their pain-relieving, mood-enhancing, and stress-reducing effects. There are several types of opioid receptors, including **mu**, **delta**, and **kappa** receptors, and genetic variations can influence the density and sensitivity of these receptors.

Opioid Receptor Gene Variations

OPRM1

- **Higher Sensitivity to Endorphins**: Individuals with certain genetic variations may experience stronger endorphin effects, leading to greater emotional resilience, enhanced mood regulation, and a lower sensitivity to pain.
- **Lower Sensitivity to Endorphins**: Conversely, some genetic variants result in lower receptor sensitivity, which may reduce the effectiveness of endorphins in pain management and mood enhancement. These individuals may need to adopt additional strategies to optimize endorphin production or rely on external sources, such as exercise or mindfulness, to boost their endorphin levels.

Endorphin Receptor Binding Efficiency

binding efficiency

How Genetics Influence Endorphin Production and Sensitivity

Genetic factors not only affect receptor sensitivity but also influence the **synthesis and release** of endorphins. Some individuals are genetically predisposed to produce higher amounts of endorphins, while others may produce lower amounts. These variations are primarily determined by genes that regulate the activity of enzymes and pathways involved in endorphin synthesis, such as the **pro-opiomelanocortin (POMC)** gene, which is responsible for producing precursor proteins that are eventually converted into endorphins.

1. **POMC Gene Variations**

   The POMC gene plays a pivotal role in endorphin synthesis. Mutations or variations in this gene can result in higher or lower levels of endorphins being produced in the body. For instance, certain genetic variations may enhance the expression of the POMC gene, leading to an increased production of endorphins, while other mutations may reduce their synthesis.

2. **Endorphin Production in Response to Stress**

   Stress is a major trigger for endorphin release, and genetic factors influence how effectively endorphins are produced in response to stress. Individuals with certain genetic profiles may have a more robust endorphin response to stress, providing them with enhanced coping mechanisms and emotional resilience. On the other hand, people with genetic predispositions for lower endorphin production during stress may experience higher levels of anxiety or depression in response to stressful events.

Personalized Approaches to Endorphin Optimization

Understanding your genetic predisposition to endorphin production and receptor sensitivity can help you develop a personalized approach to optimizing your endorphin levels. By leveraging genetic insights, individuals can create a tailored strategy that maximizes their natural ability to produce and utilize endorphins for pain relief, emotional well-being, and stress management.

Genetic Testing for Endorphin Sensitivity

- **Higher Sensitivity**: If you have higher receptor sensitivity or endorphin production, practices like moderate exercise, meditation, and social engagement may be sufficient to maintain high endorphin levels.

- **Lower Sensitivity**: If you have lower sensitivity or endorphin production, you may benefit from more intense exercise routines, higher-intensity social interactions, or even pharmacological interventions to enhance endorphin levels.

Optimizing Endorphin Synthesis Through Lifestyle Adjustments

- **Exercise**: Physical activity is one of the most effective ways to stimulate endorphin release, regardless of genetic variations. High-intensity interval training (HIIT) or endurance exercise like running, cycling, or swimming can boost endorphin levels significantly, even in individuals with less responsive endorphin pathways.

- **Mindfulness and Meditation**: Regular meditation practices can enhance endorphin production by reducing stress and promoting emotional well-being. Mindfulness-based stress reduction (MBSR) has been shown to increase both serotonin and endorphin levels.

- **Social Engagement**: Positive social interactions, such as spending time with loved ones or participating in group activities, can stimulate endorphin release and promote a sense of happiness and connectedness.

Diet and Nutrition

tryptophan

omega-3 fatty acids

antioxidants

- **Tryptophan-Rich Foods**: Tryptophan is an amino acid found in foods like turkey, eggs, and cheese. It is a precursor to serotonin, which works in conjunction with endorphins to improve mood.

- **Omega-3 Fatty Acids**: Found in fatty fish, flax seeds, and walnuts, omega-3s support brain function and enhance emotional health, which indirectly supports optimal endorphin production.

- **Dark Chocolate**: The compounds in dark chocolate, especially **flavonoids**, are known to trigger endorphin release and promote feelings of pleasure.

Genetics and Endorphin-Based Therapies

For those with genetic variations that result in low endorphin production or sensitivity, there are several therapeutic options to explore. In addition to lifestyle changes, pharmacological treatments can help enhance endorphin activity.

1. **Opioid Medications**

   Although opioid medications have been associated with addiction and abuse, they are designed to activate the brain's opioid receptors and mimic the effects of natural endorphins. In certain cases, individuals with low endorphin production may benefit from opioid-based medications, though this approach should be used with caution and under medical supervision.

2. **Non-Pharmacological Alternatives**

   For individuals with genetic predispositions to low endorphin levels, non-pharmacological interventions can also be effective. These include therapies like **electroconvulsive therapy (ECT)**, which has been shown to increase the release of neurotransmitters, including endorphins, in the brain. Additionally, transcranial magnetic stimulation (TMS) and neurofeedback may offer promising alternatives for improving endorphin function.

Conclusion

Understanding the genetic factors that influence endorphin production and receptor sensitivity can provide valuable insights into personalizing strategies for optimizing well-being. Whether through exercise, mindfulness, diet, or pharmacological interventions, there are numerous ways to enhance endorphin levels and ensure a higher quality of life. By recognizing the role of genetics in shaping endorphin function, individuals can take proactive steps to manage their endorphin system and improve both their physical and emotional health.

Chapter 16: Stress, Endorphins, and the Immune System

Stress is an inevitable part of life, but how our bodies respond to stress can significantly impact our health. Chronic stress can weaken the immune system, disrupt normal bodily functions, and increase the risk of various diseases. On the other hand, managing stress effectively can bolster the immune system, improve mood, and enhance overall health. Endorphins, the body's natural "feel-good" chemicals, play a crucial role in both coping with stress and strengthening immune function. In this chapter, we explore the relationship between stress, endorphins, and immune function, and how enhancing endorphin production can boost the immune system and improve your ability to handle stress.

The Interaction Between Endorphins and Immune Function

Endorphins are known for their role in regulating pain, mood, and stress, but they also have an important impact on immune function. Research has shown that endorphins influence the body's immune response by enhancing the activity of immune cells, improving the body's ability to fight infections, and promoting overall health.

Endorphins and Immune Activation

T-cells

B-cells

natural killer (NK) cells

- **T-cells and Endorphins**: T-cells are essential for identifying and attacking infected cells. Research has shown that endorphins can increase the number of active T-cells in the body, improving the body's ability to respond to pathogens.

- **B-cells and Antibody Production**: B-cells are responsible for producing antibodies that neutralize harmful substances. Endorphins can stimulate B-cell activity, enhancing the body's immune response.

- **Natural Killer (NK) Cells**: NK cells play a key role in the body's defense against cancer and viral infections. Studies have shown that endorphins increase the activity of NK cells, thereby improving the body's ability to target and destroy abnormal or infected cells.

Endorphins and Stress Modulation

cortisol

- **Cortisol Regulation**: Studies have shown that endorphins can help regulate cortisol levels, reducing the harmful effects of chronic stress. This is particularly important because prolonged exposure to high cortisol levels is associated with immune suppression, inflammation, and a higher susceptibility to illness.
- **Reducing Inflammation**: Chronic inflammation is another consequence of prolonged stress. Endorphins help to reduce inflammation by modulating the activity of cytokines, proteins that regulate immune responses and inflammation. By decreasing inflammation, endorphins help reduce the risk of chronic diseases such as cardiovascular disease, diabetes, and autoimmune disorders.

Endorphins and Stress Relief

dopamine

serotonin

- **Euphoria and Relaxation**: The "rush" of euphoria that accompanies the release of endorphins is a key component of stress relief. This feeling of pleasure can act as a natural antidote to stress, reducing anxiety and promoting relaxation.

- **Coping Mechanisms**: When endorphins are present in higher levels, individuals tend to develop better coping mechanisms for stress, enabling them to handle challenging situations with greater resilience.

Enhancing Endorphin Production to Boost Immunity

Given the profound impact endorphins have on immune function and stress modulation, there are several ways to enhance endorphin production and bolster the immune system. By adopting endorphin-boosting strategies, you can improve your body's resilience to stress and enhance your immune response.

Physical Activity

- **Aerobic Exercise**: Activities such as running, cycling, swimming, and dancing increase endorphin production and promote immune function. Aerobic exercise also reduces cortisol levels and enhances circulation, which further supports immune health.

- **Strength Training**: Resistance training, such as weightlifting or bodyweight exercises, can also boost endorphins and promote muscle strength and overall health. Combining both aerobic and strength exercises provides the best overall benefits for both endorphin production and immune support.

- **Moderate-Intensity Exercise**: Engaging in moderate-intensity exercise, such as brisk walking or hiking, is especially beneficial for those looking to optimize endorphin levels without overstressing the body.

Mindfulness and Relaxation Techniques

- **Meditation**: Mindfulness meditation helps lower cortisol levels and enhances endorphin release by promoting a state of relaxation and focus. Studies have shown that even short periods of meditation can improve immune function by increasing endorphin levels.

- **Deep Breathing**: Slow, deep breathing exercises, particularly those focused on abdominal breathing, can activate the vagus nerve and stimulate endorphin release. These techniques help counteract the effects of stress and improve overall well-being.

- **Yoga**: Yoga, with its combination of physical postures, controlled breathing, and relaxation techniques, is a highly effective practice for enhancing both endorphin levels and immune function.

Social Interaction and Laughter

- **Laughter**: Laughter is a powerful endorphin booster. Studies have shown that laughter not only improves mood but also enhances immune function by increasing endorphin levels. Activities like laughter yoga, watching comedies, or simply sharing jokes with loved ones can promote endorphin production and reduce stress.

- **Positive Social Interactions**: Acts of kindness, hugging, and engaging in meaningful conversations with others also stimulate endorphin release and reduce stress. These interactions create a sense of connection and safety, enhancing emotional well-being and boosting immune function.

Nutritional Support for Endorphin Production

- **Amino Acids**: Foods rich in tryptophan, such as turkey, eggs, and nuts, help the body produce serotonin, which works synergistically with endorphins to enhance mood and reduce stress.

- **Omega-3 Fatty Acids**: Found in fatty fish, flaxseeds, and walnuts, omega-3 fatty acids are known to support brain health and immune function, as well as improve emotional well-being.

- **Vitamin C**: Vitamin C is crucial for immune function and can also help reduce cortisol levels in the body, thereby indirectly enhancing endorphin production.

Conclusion

Endorphins are essential for managing stress, supporting immune function, and improving overall health. By enhancing endorphin production through exercise, relaxation techniques, social interaction, and proper nutrition, you can significantly boost your body's ability to cope with stress and enhance its immune defenses. Integrating these strategies into your daily life can help you achieve a balanced and resilient body that is better equipped to handle the challenges of life, promoting both long-term physical and mental well-being.

Chapter 17: Chronic Conditions and Endorphin Dysfunction

Chronic conditions, including both physical and mental health disorders, can often result in disruptions to normal biological processes, including the synthesis, release, and regulation of endorphins. Endorphin dysfunction can exacerbate symptoms, hinder recovery, and make it more difficult for individuals to cope with their condition. Understanding how endorphins play a role in these chronic conditions and how imbalances in endorphin levels can lead to further complications is critical for managing long-term health. In this chapter, we explore the role of endorphin deficiency in chronic illnesses, the concept of endorphin resistance, and the potential treatments and therapies that can help restore balance and improve overall well-being.

Endorphin Deficiency and Its Role in Chronic Illnesses

Endorphins, often described as the body's natural painkillers, are responsible for promoting feelings of euphoria, improving mood, and reducing pain and stress. A deficiency in endorphins can manifest in both physical and mental health disorders, leading to greater susceptibility to chronic illnesses. Some of the chronic conditions that are linked to endorphin deficiencies include:

Chronic Pain Conditions

- **Fibromyalgia and Endorphin Deficiency**: Fibromyalgia is a common chronic pain condition that causes widespread muscle pain, fatigue, and cognitive issues. Research has shown that individuals with fibromyalgia may have impaired endorphin production, which contributes to the heightened pain and discomfort they experience. Increasing endorphin levels through lifestyle changes such as exercise, diet, and stress management can significantly help alleviate some of the symptoms of fibromyalgia.

- **Osteoarthritis and Joint Pain**: Osteoarthritis is a degenerative joint disease that causes pain, stiffness, and swelling in the joints. Endorphins are essential for modulating pain, and a deficiency can make osteoarthritis symptoms worse. Endorphin-based therapies or strategies, such as exercise or acupuncture, can be used to stimulate the body's natural pain-relieving mechanisms.

Mental Health Conditions

- **Depression and Low Endorphin Levels**: Studies have shown that individuals with depression often have lower levels of endorphins, which may contribute to their low mood, lack of motivation, and inability to experience pleasure. Rebalancing endorphins through exercise, nutrition, or therapy can help individuals suffering from depression regain a sense of well-being.

- **Anxiety and Stress**: Anxiety disorders are frequently associated with altered endorphin levels. In some cases, low endorphins may contribute to heightened anxiety and stress. Restoring endorphin function through regular physical activity or relaxation techniques can help alleviate symptoms of anxiety and promote a more balanced emotional state.

Endorphin Resistance and Its Health Implications

Endorphin resistance is a concept similar to insulin resistance, where the body's cells become less responsive to the signals of endorphins. This can occur in individuals who experience chronic stress, substance abuse, or prolonged exposure to pain. When endorphin receptors become desensitized, the body may require higher levels of endorphins to achieve the same effect, leading to a cycle of imbalance.

1. **Chronic Stress and Endorphin Resistance**

   Prolonged exposure to stress leads to elevated cortisol levels, which can impair the function of endorphin receptors. This phenomenon contributes to endorphin resistance, as the body becomes less able to experience the stress-relieving and pain-reducing effects of endorphins. Individuals experiencing this resistance may have a decreased ability to cope with stress, leading to emotional and physical strain.

2. **Substance Abuse and Endorphin Dysfunction**

   The use of substances such as opioids, alcohol, or other drugs can interfere with the body's natural endorphin production. Opioids, for example, bind directly to opioid receptors in the brain and artificially increase endorphin levels, but with repeated use, the body's own ability to produce endorphins can diminish. This leads to a cycle where individuals rely more on external substances for pain relief or emotional regulation, exacerbating endorphin dysfunction.

3. **Chronic Pain and Endorphin Resistance**

   In cases of chronic pain, such as with migraines or post-surgical pain, endorphin resistance can develop as the body becomes desensitized to pain-relieving endorphins. As the pain persists, the body's ability to generate its own natural

Treatments and Therapies for Endorphin Imbalance

   painkillers diminishes, requiring external interventions like pain medications.

Restoring proper endorphin function in individuals experiencing deficiency or resistance is essential for managing chronic conditions and improving overall health. Several strategies can help boost endorphin production, including lifestyle modifications, alternative therapies, and, in some cases, medical treatments.

1. **Exercise and Physical Activity**

   Regular physical activity is one of the most effective ways to increase endorphin production. Exercise stimulates the release of endorphins, improving mood and helping to alleviate pain. Engaging in activities such as aerobic exercises, resistance training, yoga, or even moderate walking can stimulate endorphin release and help break the cycle of endorphin resistance. For chronic pain and depression, targeted exercises can enhance overall well-being by improving both physical and mental health.

2. **Nutrition and Diet**

   Nutrition plays a key role in supporting endorphin production. Foods rich in amino acids such as **tryptophan, phenylalanine**, and **tyrosine** help the body synthesize endorphins. Foods like turkey, eggs, nuts, and seeds are great sources of these essential amino acids. Omega-3 fatty acids, found in fish and flaxseeds, also support brain health and endorphin production, while foods high in antioxidants help reduce inflammation that may block endorphin activity.

3. Psychotherapy and Counseling

Cognitive Behavioral Therapy (CBT) and other forms of counseling can help individuals address underlying emotional issues that contribute to stress, anxiety, and depression. By managing negative thought patterns and learning coping skills, individuals can promote healthy endorphin function and improve their response to stress.

4. Alternative Therapies

Practices like acupuncture, massage therapy, and chiropractic care have been shown to stimulate endorphin release. Acupuncture, for example, activates specific points on the body that trigger the production of endorphins, reducing pain and promoting relaxation. Similarly, massage therapy can enhance circulation and increase endorphin levels, providing both mental and physical benefits.

5. Pharmacological Approaches

In some cases, medications that promote the release of endorphins or that block the activity of substances inhibiting endorphin function may be prescribed. For example, certain antidepressants and pain-relieving medications can stimulate endorphin release to help manage symptoms of depression, chronic pain, and anxiety. However, these should be used cautiously and under the supervision of a healthcare provider.

Conclusion

Endorphin dysfunction, including deficiency and resistance, is a significant factor in many chronic health conditions. By understanding the role of endorphins in both mental and physical health, we can implement strategies to address endorphin imbalances and optimize their production. Exercise, nutrition, alternative therapies, and psychotherapy all offer effective ways to restore endorphin function and improve quality of life. By taking a comprehensive approach to boosting endorphin levels, individuals can reduce the impact of chronic illness, enhance their resilience to stress, and enjoy better physical and mental health.

Chapter 18: Enhancing Endorphin Production Naturally

Endorphins play an essential role in both physical and mental well-being. From pain relief to mood elevation, these natural peptides are critical for maintaining balance in the body. However, various factors can impair endorphin production, leading to feelings of stress, anxiety, pain, or depression. Fortunately, there are numerous natural ways to enhance the body's production of endorphins. In this chapter, we will explore lifestyle changes, including exercise, diet, sleep, and mind-body practices, that can help boost endorphin levels and optimize overall well-being.

1. Exercise: The Endorphin Powerhouse

Regular physical activity is one of the most effective natural ways to stimulate endorphin production. When you exercise, the body releases endorphins, which help to improve mood, reduce stress, and even alleviate pain. Exercise-induced endorphin release is often referred to as the "runner's high," a feeling of euphoria and well-being that many athletes experience during or after intense physical activity.

Key Exercise Types for Endorphin Optimization:

- **Aerobic Exercise:** Activities like running, cycling, swimming, and dancing are particularly effective at boosting endorphins. These types of exercises elevate heart rate, which stimulates the release of endorphins and other neurotransmitters that enhance mood and energy levels.
- **Strength Training:** Resistance training, such as weightlifting, can also increase endorphin levels by placing physical stress on muscles, encouraging a response from the brain that enhances the production of endorphins.
- **Yoga and Pilates:** Low-impact activities like yoga and Pilates engage the body and mind, helping reduce stress, promote flexibility, and stimulate endorphin release through deep, mindful movement and breathing.

Exercise Guidelines:

- **Duration and Consistency:** Aim for at least 30 minutes of moderate exercise, such as brisk walking or cycling, three to five times a week. However, even shorter sessions of high-intensity interval training (HIIT) have been shown to effectively boost endorphins.

- **Mix It Up:** Engage in a variety of exercise types to keep things interesting and challenge different parts of the body. Switching between cardio, strength training, and flexibility exercises ensures balanced endorphin production and prevents workout burnout.

2. Diet: Nourishing the Endorphin Pathways

Nutrition plays a vital role in supporting the body's natural ability to produce endorphins. Foods rich in certain nutrients can directly impact endorphin production, while a balanced diet overall ensures the body has the right building blocks for efficient endorphin synthesis.

Key Nutrients for Endorphin Production:

- **Amino Acids:** Endorphins are synthesized from amino acids, which are the building blocks of proteins. Foods rich in amino acids, particularly tryptophan, phenylalanine, and tyrosine, can help the body produce more endorphins. These amino acids are found in high-protein foods such as turkey, eggs, fish, nuts, seeds, and tofu.

- **Healthy Fats:** Omega-3 fatty acids, found in fatty fish (like salmon and mackerel), flaxseeds, and walnuts, support brain health and can enhance endorphin receptor activity.

- **Dark Chocolate:** Dark chocolate, particularly varieties with at least 70% cocoa, contains compounds that trigger the brain's reward system, promoting the release of endorphins.

- **B Vitamins:** B-vitamins, especially B6, B9 (folate), and B12, are involved in neurotransmitter function and are crucial for the synthesis of endorphins. Leafy greens, beans, whole grains, and fortified cereals are excellent sources of these vitamins.

Dietary Tips:

- **Balanced Meals:** Aim to include protein, healthy fats, and fiber in every meal to maintain steady energy levels and prevent blood sugar crashes, which can negatively affect mood and endorphin production.

- **Hydration:** Staying hydrated is important for optimal cellular function. Dehydration can hinder the production of endorphins and other neurotransmitters, so aim for at least 8 cups (2 liters) of water per day, and more if you are physically active.

3. Sleep: The Secret to Restorative Endorphin Production

Sleep is a cornerstone of mental and physical health, and it plays an integral role in maintaining healthy endorphin levels. During sleep, the body undergoes repair and regeneration, including the replenishment of neurotransmitters like endorphins. Lack of quality sleep can disrupt endorphin synthesis, leading to fatigue, irritability, and impaired stress response.

The Role of Sleep in Endorphin Production:

- **Deep Sleep and REM Sleep:** Endorphin production is closely linked to the sleep cycle. The restorative phases of sleep, particularly REM (Rapid Eye Movement) sleep, are when the body experiences the highest release of endorphins. This period of deep sleep is essential for memory consolidation, emotional regulation, and overall brain function.

- **Impact of Sleep Deprivation:** Chronic sleep deprivation reduces the body's ability to produce and regulate endorphins, which may contribute to the heightened stress response, low mood, and increased pain sensitivity.

Sleep Optimization Strategies:

- **Sleep Hygiene:** Maintain a consistent sleep schedule by going to bed and waking up at the same time each day. Create a calming bedtime routine that signals to your body it's time to wind down (e.g., reading, light stretching, or meditation).

- **Limit Screen Time:** Avoid screens (phones, computers, TVs) for at least an hour before bedtime, as the blue light emitted by electronic devices can disrupt the production of melatonin, a hormone that regulates sleep.

- **Create a Sleep-Friendly Environment:** Ensure your bedroom is cool, dark, and quiet. Use blackout curtains, a white noise machine, or earplugs if necessary to create an ideal sleep environment.

4. Mind-Body Practices: Cultivating Endorphins through Focused Relaxation

Mind-body practices such as meditation, mindfulness, and deep breathing exercises can help activate endorphin production by encouraging relaxation and reducing stress.

Mindfulness Meditation: Mindfulness meditation, which involves focusing on the present moment and cultivating awareness, has been shown to promote the release of endorphins. Studies have demonstrated that even brief periods of mindfulness can increase endorphin levels while simultaneously lowering cortisol, the stress hormone.

Deep Breathing Techniques: Breathing exercises, such as deep abdominal breathing or box breathing, encourage the body's parasympathetic nervous system (the "rest and digest" system), which triggers endorphin release and reduces the harmful effects of stress. These techniques help slow the heart rate, lower blood pressure, and induce relaxation.

Yoga and Tai Chi: Both yoga and Tai Chi combine physical postures, breath control, and meditation, which together stimulate endorphin production and reduce stress. The slow, deliberate movements in these practices activate the body's natural relaxation response, promoting endorphin release and reducing physical tension.

Integrating Mind-Body Practices: Incorporating daily or weekly sessions of meditation, deep breathing, or yoga into your routine can boost overall endorphin levels and help maintain mental and emotional balance. Even just 10-15 minutes of mindful breathing can be enough to trigger endorphin release and improve your emotional state.

5. Conclusion: Enhancing Endorphins for Lifelong Well-Being

Enhancing endorphin production naturally is an ongoing process that requires attention to lifestyle choices, including exercise, nutrition, sleep, and mind-body practices. By adopting habits that prioritize physical movement, nourishing foods, restorative sleep, and mindfulness, you can support optimal endorphin levels, resulting in a healthier, more resilient body and mind.

By integrating these strategies into your daily routine, you will not only enhance your endorphin production but also cultivate a sustainable foundation for well-being. A well-balanced approach to living—where endorphin optimization is prioritized—can promote physical vitality, mental clarity, emotional balance, and a deep sense of joy throughout your life.

Chapter 19: Endorphins in Pain Management and Healing

Pain is one of the most common reasons people seek medical attention, and its impact on quality of life can be profound. Whether it's chronic pain from conditions like arthritis or temporary pain from an injury or surgery, managing pain effectively is a central component of health and well-being. Endorphins, known for their natural analgesic properties, play a critical role in modulating pain perception and promoting healing. In this chapter, we will explore the mechanisms by which endorphins reduce pain, their role in the body's healing processes, and how to harness endorphin production as part of a pain management strategy.

Endorphins as Natural Painkillers

Endorphins are often referred to as the body's **natural painkillers** because they have the ability to reduce the sensation of pain and promote a sense of well-being. This ability is attributed to their action on the **opioid receptors** in the brain and nervous system, which are the same receptors targeted by opioid medications. When endorphins bind to these receptors, they inhibit the transmission of pain signals, dampening the sensation of pain and producing an analgesic effect.

Research has shown that endorphins are particularly effective in alleviating both **acute** and **chronic pain**. Acute pain, such as that caused by an injury, triggers the release of endorphins as part of the body's natural response to stress and injury. This helps to minimize discomfort and prevent long-term harm. Chronic pain conditions, such as fibromyalgia or lower back pain, may result from a deficiency in endorphin production or a dysfunction in the body's opioid receptor system, which contributes to the persistence of pain signals.

How Endorphins Facilitate Healing

In addition to their pain-relieving properties, endorphins also play a vital role in the healing process. When endorphin levels are elevated, the body enters a state of **increased relaxation and repair**, which accelerates tissue healing. This is particularly important after injury or surgery when the body is working to repair damaged tissues.

Endorphins stimulate the **immune system**, enhancing the body's ability to fight off infections and reduce inflammation, which can also accelerate the healing of injuries. By promoting a sense of well-being and reducing the perception of pain, endorphins also contribute to psychological resilience, which is essential for recovery. Individuals who experience less emotional distress and pain are better able to engage in rehabilitation exercises and adopt the behaviors necessary for full recovery.

Endorphin-Boosting Strategies for Pain Relief

To maximize the natural pain-relieving effects of endorphins, certain lifestyle strategies and practices can be employed. These strategies can be used in conjunction with medical treatments to enhance pain management and promote faster healing.

1. **Exercise**: Regular physical activity is one of the most effective ways to trigger endorphin release. Aerobic exercises, such as running, cycling, or swimming, as well as strength training, have been shown to boost endorphin production. Even moderate exercise, such as walking or yoga, can stimulate endorphin release and help reduce pain levels.

2. **Acupuncture and Acupressure**: These traditional healing practices involve stimulating specific points on the body, which can trigger endorphin release and provide relief from chronic pain.

3. **Massage**: Therapeutic massage has been shown to increase endorphin levels and reduce muscle tension, providing natural pain relief. Regular massages can help manage both acute and chronic pain conditions, including tension headaches and lower back pain.

4. **Mind-Body Practices**: Mindfulness, meditation, and deep breathing exercises can reduce stress and increase endorphin production. Practices such as **Tai Chi** and **Qigong** integrate movement and focused breathing to promote both physical and mental well-being, enhancing the body's natural healing abilities.

5. **Exposure to Sunlight**: Sunlight is a natural stimulant for endorphin production. Ensuring adequate exposure to sunlight, while maintaining safe sun practices, can boost mood and pain tolerance.

6. **Diet and Nutrition**: Certain foods, such as dark chocolate, spicy foods, and foods rich in omega-3 fatty acids, have been shown to enhance endorphin levels. A balanced diet rich in nutrients also supports overall health and well-being, facilitating the body's ability to heal.

7. **Laughter**: Laughter is a natural trigger for endorphin release. Engaging in activities that promote laughter, such as watching a funny movie, spending time with friends, or practicing laughter yoga, can boost mood and provide natural pain relief.

When Endorphin Dysfunction Affects Pain Management

For some individuals, endorphin production may be insufficient or dysfunctional, which can impair the body's ability to manage pain effectively. This dysfunction may occur due to genetic factors, chronic stress, mental health conditions, or the use of substances that interfere with the body's opioid receptors, such as opioids and alcohol.

In cases of **endorphin deficiency**, individuals may experience heightened sensitivity to pain, difficulty coping with stress, and an increased likelihood of developing chronic pain conditions. In such cases, it may be necessary to explore medical interventions in combination with lifestyle changes to restore endorphin balance and improve pain management.

For individuals with **chronic pain conditions**, therapies such as **endorphin-enhancing medications**, **cognitive-behavioral therapy (CBT)** for pain management, or even **biofeedback** techniques may be helpful in restoring normal endorphin function and providing relief.

The Role of Endorphins in Emotional Pain

Emotional pain, such as grief, depression, and anxiety, can feel just as intense as physical pain. Endorphins are essential for regulating emotions and maintaining psychological resilience. They play a crucial role in managing the distress that comes with emotional suffering by providing a natural sense of comfort and relief.

By boosting endorphin levels through social engagement, exercise, and positive lifestyle habits, individuals can reduce the emotional burden of pain and build a more resilient mindset. Combining physical pain management with emotional support, including **therapy**, **support groups**, and **mindfulness practices**, can create a holistic approach to dealing with pain in all its forms.

Conclusion: Harnessing Endorphins for Pain Relief

Endorphins are a powerful, natural tool in the fight against both physical and emotional pain. By understanding how these chemicals work, we can better harness their benefits to support healing, manage chronic conditions, and improve overall well-being. By incorporating endorphin-boosting practices into your daily life—whether through exercise, diet, mindfulness, or alternative therapies—you can enhance your body's ability to manage pain and heal more effectively, empowering yourself to live a healthier, pain-free life.

Chapter 20: Endorphins and Social Connection: The Power of Human Interaction

Human beings are inherently social creatures, and our relationships with others play a critical role in our mental, emotional, and physical well-being. One of the key biochemical players in fostering social bonds and enhancing positive emotions is **endorphins**. Whether it's through touch, laughter, or simply sharing experiences, endorphins are essential for creating feelings of connection, trust, and happiness. In this chapter, we will explore how endorphins are released during social interactions, the importance of social bonding for mental health, and how you can enhance your endorphin levels through meaningful relationships and social engagement.

The Role of Endorphins in Social Bonding

Endorphins are involved in forming and maintaining social bonds in multiple ways. One of the most well-documented mechanisms...

Chapter 21: Lifestyle Practices for Sustaining Optimal Endorphin Levels

In our fast-paced world, it can be challenging to maintain a consistent sense of well-being, especially when life's stressors take a toll on our mental and physical health. However, small, consistent lifestyle changes can have a profound impact on sustaining optimal endorphin production. In this chapter, we will explore various daily habits, routines, and practices that can support long-term endorphin balance. From simple adjustments to diet and sleep...

Chapter 22: The Future of Endorphin Research: Unlocking New Frontiers

As we continue to understand more about the biochemical pathways, functions, and impacts of **endorphins**, the potential for future breakthroughs in health and medicine grows exponentially. Researchers are delving deeper into how endorphins influence a variety of conditions, from mental health disorders to chronic pain and autoimmune diseases. In this chapter, we will explore the current state of endorphin research, exciting developments on the horizon, and how these discoveries may lead to new therapies, treatments, and even enhancements in human performance and longevity.

Current Research on Endorphins: A Broadening Understanding

Endorphins have long been understood as the body's natural response to pain and stress, but modern research is expanding their role in ways that were previously unimagined. Studies are now exploring how endorphins interact with the body's **immune system, neurotransmitters**, and even **genetic expressions**, revealing a more complex relationship with health and disease. Researchers are also investigating the ways endorphins contribute to cognitive function, emotional resilience, and the overall aging process.

1. **Endorphins and Mental Health**: One exciting area of research involves the role of endorphins in mental health conditions such as depression, anxiety, and PTSD. Studies suggest that endorphin imbalances may contribute to mood disorders, and enhancing endorphin production through specific treatments could offer an alternative to traditional pharmacological interventions.

2. **Endorphins and Chronic Pain Management**: Chronic pain has been a significant focus of endorphin research, with particular attention given to understanding how endorphin deficiencies or resistance can make pain more persistent and difficult to treat. New therapies, such as targeted **gene therapies** or **neurostimulation techniques**, aim to enhance natural endorphin production to provide long-term pain relief without the risks associated with opioids.

3. **Endorphins and Autoimmune Diseases**: Autoimmune diseases, where the body's immune system attacks its own tissues, are another area where endorphins may have therapeutic potential. Emerging research suggests that endorphins could play a role in modulating immune responses, helping to reduce inflammation and improve symptoms in diseases like rheumatoid arthritis, lupus, and multiple sclerosis.

The Role of Technology in Endorphin Optimization

Advancements in **neurotechnology** and **genomics** may revolutionize how we approach endorphin production in the future. Cutting-edge devices that monitor neural activity and hormones, along with genetic testing, could enable more personalized approaches to enhancing endorphin levels. For example, genetic profiling could reveal whether a person has genetic predispositions that make them more or less susceptible to low endorphin levels, guiding lifestyle and treatment choices tailored to their unique needs.

1. **Wearable Devices**: Wearables designed to track both physical activity and emotional states could soon be equipped with real-time biofeedback systems that encourage activities known to boost endorphins. These devices could alert wearers when they are under stress or low on energy, prompting them to engage in activities like deep breathing, exercise, or social interactions to restore balance.

2. **Neurostimulation**: Techniques such as **transcranial magnetic stimulation (TMS)** or **deep brain stimulation (DBS)** are currently being explored for their potential to boost endorphin activity in the brain, offering hope for individuals with depression, anxiety, and chronic pain who have not responded well to conventional treatments.

3. **Gene Therapy**: As our understanding of genetics improves, gene therapy may offer a future solution to endorphin deficiencies. By editing the genes responsible for endorphin production or receptor sensitivity, we could potentially help individuals with conditions like chronic pain, addiction, or mood disorders by enhancing their natural endorphin response.

Ethical Considerations in Endorphin Enhancement

As with any new technology or medical treatment, the ability to enhance or regulate endorphin production raises ethical questions. For example, should endorphin-boosting treatments be available for people who simply want to enhance their mood or performance? Where should the line be drawn between natural biological processes and artificially induced states of well-being? Additionally, as we consider genetic manipulation, what are the long-term effects on an individual's health and the broader population?

The ethical implications of endorphin research and enhancement will require careful consideration, particularly as the ability to manipulate neurochemistry becomes more refined. Balancing the benefits of enhanced well-being with the risks of over-reliance on technological or pharmaceutical interventions will be a critical challenge for future healthcare systems.

Conclusion: The Promise of Endorphin Science

The future of endorphin research holds immense promise, not only for treating existing health conditions but also for enhancing human well-being in unprecedented ways. As we continue to unravel the complexities of these powerful molecules, we are likely to see new treatments, therapies, and interventions that will transform how we manage pain, mental health, and chronic conditions. With this deeper understanding, we may unlock new ways to optimize our natural biochemistry and harness the full potential of our body's endorphin system to lead healthier, more fulfilling lives.

Ultimately, mastering the science of endorphins is not just about understanding their chemical pathways—it's about harnessing this knowledge to cultivate a more balanced, resilient, and vibrant life for individuals across the world. The future of endorphin science is just beginning, and it holds the potential to revolutionize our approach to health, wellness, and human flourishing.

Chapter 23: Integrating Endorphin Optimization Into Everyday Life

Mastering the synthesis, production, and availability of endorphins requires a holistic approach. It's not enough to rely on one strategy; to truly harness the power of endorphins, we need to integrate endorphin-boosting habits into our daily lives. This chapter will provide actionable steps and strategies that anyone can incorporate into their routine to optimize endorphin levels for sustained physical and mental well-being.

1. Create a Balanced Routine: Consistency is Key

One of the most important factors in optimizing endorphin production is consistency. Our bodies thrive on regular routines that promote balance and health. This means incorporating activities that stimulate endorphin release into your daily schedule, whether that's exercise, mindfulness, social interactions, or even creative expression.

- **Exercise:** Aim for at least 30 minutes of moderate aerobic exercise most days of the week. Activities like running, swimming, cycling, or dancing are particularly effective at triggering endorphin release. For those with limited time, high-intensity interval training (HIIT) is another powerful strategy for endorphin production.

- **Mindfulness Practices:** Daily meditation or mindfulness practices, even if it's just for 10 minutes, can significantly boost endorphin levels while reducing stress and promoting a sense of well-being.

- **Sleep:** A solid sleep routine is paramount for endorphin production. Aim for 7–9 hours of sleep each night to allow for proper recovery and regulation of neurotransmitters.

- **Nutrition:** Eating nutrient-dense foods that support brain health—such as omega-3-rich foods, antioxidants, and lean proteins—can help optimize the body's natural ability to produce and regulate endorphins.

2. Foster Social Connection and Meaningful Relationships

Social interactions are a critical aspect of mental and emotional well-being. Positive social relationships—whether they're with friends, family, or colleagues—activate endorphins, creating feelings of happiness and fulfillment.

- **Quality Time:** Make time each week to engage in meaningful interactions, whether that's a conversation with a loved one or participating in group activities that promote a sense of belonging.
- **Laughter and Play:** Humor is one of the most powerful triggers for endorphin release. Try to find moments in your day to laugh, whether that's watching a funny movie, engaging in playful activities, or simply being around people who make you smile.
- **Acts of Kindness:** Doing something kind for others has been shown to release endorphins and promote feelings of happiness and connectedness. Whether through volunteering, random acts of kindness, or just helping someone in need, these small actions can make a significant impact on your well-being.

3. Embrace Creativity and Self-Expression

Engaging in creative pursuits or activities that allow for self-expression can also trigger the release of endorphins. This could be anything from painting or writing to playing music or crafting. The key is to immerse yourself in an activity that brings you joy and a sense of accomplishment.

- **Creative Hobbies:** Dedicate time to activities like painting, drawing, writing, or photography. These activities are not only therapeutic but can help boost your mood by promoting the release of endorphins.
- **Flow State:** Try to immerse yourself in "flow" experiences—those times when you lose yourself in an activity and feel fully engaged. Whether it's playing a sport, working on a challenging project, or performing a task you love, these moments of deep engagement are often associated with endorphin release.

4. Use Stress Management Techniques

While stress is an unavoidable part of life, how we manage it can have a significant impact on our endorphin levels. Regularly practicing stress-reduction techniques can help maintain emotional balance and prevent the negative effects of chronic stress.

- **Breathing Exercises:** Deep, diaphragmatic breathing can lower cortisol levels and stimulate the production of endorphins. Try incorporating breathing exercises into your routine, especially during moments of tension or anxiety.

- **Yoga and Tai Chi:** Both yoga and tai chi are excellent ways to reduce stress while promoting physical movement and mindfulness. These practices are known to boost endorphins and improve overall emotional resilience.

- **Nature Therapy:** Spending time outdoors, whether through hiking, gardening, or simply walking in nature, has been shown to reduce stress and increase endorphin levels. Aim to spend time outside every day, even if it's just for a short walk.

5. Cultivate a Positive Mindset

Your thoughts and attitudes play a significant role in your emotional state and can influence endorphin production. Practicing gratitude, positive self-talk, and mindfulness can help shift your mindset and enhance your emotional resilience.

- **Gratitude Practice:** Research shows that regularly focusing on what you're grateful for can enhance mood and increase endorphin levels. Keep a gratitude journal and take time each day to reflect on the positive aspects of your life.

- **Positive Self-Talk:** Challenge negative thoughts and replace them with empowering, positive affirmations. This shift in mindset can help foster emotional well-being and promote endorphin release.

- **Visualization:** Visualization techniques—such as imagining a positive outcome or mentally rehearsing a challenging task—can also stimulate endorphin production, boosting confidence and mood.

6. Optimize Your Environment for Endorphin Production

Your surroundings can have a profound impact on your mental and emotional state. Creating an environment that promotes relaxation, joy, and connection can help foster the conditions necessary for endorphin release.

- **Clutter-Free Living Space:** Organize your home or workspace to minimize stress. A tidy, well-organized environment can promote a sense of calm and well-being.

- **Nature in Your Environment:** Bring nature indoors by adding plants, natural light, or elements like water features. These additions can create a sense of tranquility and help regulate your mood.

- **Aromatherapy:** Certain scents, such as lavender, peppermint, and citrus, have been shown to positively affect mood and reduce stress, potentially boosting endorphin levels.

7. Consistency and Patience

Ultimately, the key to mastering endorphin optimization is consistency. Incorporating endorphin-boosting habits into your daily life will yield the best results when done regularly over time. While the effects may not be immediate, the cumulative impact of these practices will significantly enhance your overall physical and emotional well-being.

Chapter 24: Sustaining Endorphin Optimization in the Long-Term: Overcoming Challenges and Setbacks

Achieving a balance in endorphin production and availability is an ongoing process, one that requires dedication, consistency, and an understanding of the challenges that may arise. While the strategies discussed throughout this book are highly effective, there will inevitably be times when your endorphin levels may fluctuate due to life circumstances, stress, illness, or other external factors. In this chapter, we will discuss how to navigate these challenges, prevent setbacks, and ensure that endorphin optimization remains a sustainable part of your lifestyle over the long term.

1. The Impact of Stress and Life Events on Endorphin Production

Chronic stress is one of the most significant barriers to optimal endorphin production. Stress reduces the efficiency of the **hypothalamic-pituitary-adrenal (HPA) axis**, which plays a role in endorphin release, and leads to **endorphin depletion** over time. Major life events—such as a loss, career challenges, or relationship issues—can significantly disrupt your mood, making it harder to maintain consistent endorphin-enhancing practices.

To counteract these disruptions, it's essential to integrate **coping strategies** that mitigate stress. Deep breathing, progressive muscle relaxation, and mindfulness techniques can help rebalance your neurochemical systems and ensure endorphin levels remain adequate even during difficult times. It's also important to acknowledge that recovery is not always linear, and during periods of emotional or physical hardship, **patience and self-compassion** are key to maintaining long-term endorphin optimization.

2. Dealing with Plateaus in Endorphin Levels

When first implementing new strategies to boost endorphins, the effects are often rapid and noticeable. However, over time, it's common for the body to reach a **plateau**, where it may feel as if the benefits have diminished. This can be discouraging, but it's important to recognize that your body is simply adjusting to the changes. To overcome plateaus, it's helpful to **introduce variety** into your endorphin-boosting activities. Switching up your exercise routine, trying new mindfulness practices, or incorporating a different social activity can provide fresh stimulation to the body's systems.

Sometimes, the solution may involve **increasing intensity** or frequency in specific areas. For example, if physical exercise has lost some of its punch in terms of endorphin release, consider increasing the intensity of your workouts or incorporating more challenging activities like HIIT (high-intensity interval training). Similarly, if meditation no longer provides the same sense of calm, try extending the duration of your practice or experimenting with different types of meditation.

3. Health Conditions and Their Impact on Endorphin Levels

Certain **health conditions**, both physical and mental, can significantly impair the body's ability to produce and regulate endorphins. Chronic pain, depression, anxiety disorders, autoimmune diseases, and hormonal imbalances all play a role in diminishing endorphin production. In these cases, it may be necessary to work in tandem with medical professionals to address the root causes of these conditions.

For those dealing with such challenges, **supportive therapies**, such as counseling, physical therapy, medication, or specialized nutritional advice, may be needed to facilitate healing and restore the body's natural ability to produce endorphins. It's also important to remember that **holistic approaches** that focus on the mind-body connection—such as yoga, acupuncture, or neurofeedback—can complement conventional treatments and provide additional relief.

4. The Importance of Sleep and Rest for Long-Term Success

While optimizing endorphin production requires active effort, it's just as crucial to allow your body to rest and regenerate. Sleep is one of the most critical components of **endocrine health** and plays a direct role in the regulation of endorphin production. Sleep deprivation not only affects mood, cognition, and energy levels, but it can also interfere with the body's ability to synthesize endorphins efficiently.

To ensure long-term success in endorphin optimization, prioritize **quality sleep**. Establishing a regular sleep schedule, limiting screen time before bed, and creating a calming bedtime routine can help improve sleep quality and support natural endorphin production.

5. Building Resilience to Overcome Long-Term Setbacks

Endorphin optimization is a journey, and like any journey, there will be ups and downs. Building **resilience** is key to maintaining your progress, even when obstacles arise. This means developing a mindset that sees challenges as opportunities for growth rather than roadblocks. Cultivating mental flexibility, learning how to adapt to changing circumstances, and setting **realistic goals** can help you overcome setbacks and stay motivated.

Additionally, **self-compassion** plays a significant role in sustaining long-term success. When you experience setbacks, it's easy to fall into a cycle of self-blame or frustration. Instead, approach these moments with kindness, acknowledge the difficulty, and remember that setbacks are simply part of the human experience. With resilience and persistence, you can continue to reap the benefits of endorphin optimization for years to come.

Conclusion

Mastering the synthesis, production, and availability of endorphins is not a one-time effort; it's an ongoing practice that requires commitment, adaptability, and awareness. By understanding the challenges that may arise and developing strategies to overcome them, you can sustain endorphin optimization in your daily life, leading to lasting physical and mental well-being. As we've explored throughout this book, the power of endorphins can profoundly influence your ability to manage stress, pain, and emotional challenges—empowering you to live a more fulfilling and vibrant life.

Chapter 25: Conclusion: Harnessing the Power of Endorphins for Lifelong Health and Well-Being

Throughout this book, we've explored the incredible role that **endorphins** play in enhancing both our physical and emotional states. From pain relief and stress reduction to mood enhancement and improved immune function, the power of endorphins cannot be overstated. By understanding the biochemistry behind their synthesis, identifying the factors that influence their production, and incorporating practical strategies for boosting endorphin levels into our daily lives, we can unlock a transformative pathway to better health, happiness, and resilience.

The Journey to Mastering Endorphins

Mastering endorphin production isn't a one-time event—it's a lifelong journey. It's about building and maintaining habits that promote well-being and resilience, allowing the body to naturally produce these powerful chemicals. Whether you're using exercise, mindfulness practices, diet, or social interactions to boost endorphins, the key is consistency. The process of improving your endorphin function doesn't require drastic changes overnight but can be achieved step by step. By incorporating the tools and strategies outlined in this book, you are well-equipped to optimize your endorphin production for enhanced health and vitality.

Empowerment Through Knowledge and Action

As with any area of health, knowledge empowers action. The more you understand the factors that influence your body's endorphin production, the more control you gain over your physical and emotional well-being. This book has provided you with a comprehensive understanding of how endorphins work, how to improve their availability, and how to integrate strategies into your life for ongoing support. But ultimately, the most important step is taking action.

The body's natural capacity for healing, mood regulation, and stress management through endorphins is immense. The key lies in **self-care**—understanding your needs, making conscious choices that promote balance, and listening to your body when it signals the need for support. By using the tools shared here, you can foster a thriving body and mind, even in the face of life's inevitable challenges.

A Holistic Approach to Health and Well-Being

Endorphins are only one part of a larger holistic approach to health. While optimizing endorphins is crucial, it is essential to remember that the body operates as an interconnected system. Factors such as sleep, nutrition, exercise, social connections, and mental health all play a role in creating balance and harmony within.

This book has outlined specific ways to enhance endorphin production, but it is important to complement these strategies with broader wellness practices. Whether through balanced nutrition, managing stress, practicing mindfulness, or fostering healthy relationships, it's the combination of these elements that creates a vibrant and sustainable path to well-being.

The Future of Endorphin Mastery

Looking ahead, we can expect even more breakthroughs in understanding the intricacies of endorphins and their role in human health. As science advances, we will gain more insight into how to fine-tune our bodies' endorphin systems, potentially leading to novel treatments and therapies for a variety of conditions, from chronic pain to mental health disorders.

With the growing knowledge of genetics, neuroscience, and biochemistry, personalized approaches to endorphin optimization may become increasingly accessible, allowing individuals to tailor strategies that best suit their unique physiology. This represents an exciting frontier in medicine and wellness—one that promises a more individualized, precise approach to improving mental and physical health.

Your Path Forward

As you conclude this book, take a moment to reflect on what you've learned and consider how you will integrate these practices into your daily life. Endorphins are powerful allies in the pursuit of a healthier, happier, and more resilient life. By understanding how they work and committing to lifestyle changes that optimize their production, you have the potential to transform your physical and mental health.

Remember that achieving optimal well-being is not a destination—it's a continuous journey of growth, self-care, and mindfulness. Stay committed to nurturing your body's natural ability to produce and regulate endorphins, and watch as your well-being flourishes.

Thank you for embarking on this journey to mastering endorphin synthesis, production, and availability. May you continue to harness their power to live your best life—now and for years to come.